Predicting Rainfall through Ensemble Machine Learning to Boost Agricultural Output

Veeraankalu Vuyyuru

Predicting Rainfall through Ensemble Machine Learning to Boost Agricultural Output

First Edition February 2024

Written by Veeraankalu Vuyyuru

CONTENTS

Content   **Page Number**

LIST OF TABLES

LIST OF FIGURES

LIST OF ABBREVIATIONS

CMIP5	:	Climate Model Inter-comparison Project Phase 5
WMO	:	World Meteorological Organization
VAE	:	Variational Auto-Encoder
DIANA	:	divisive hierarchical clustering algorithm
KNN	:	K-Nearest Neighbor
ROC	:	Receiver Operating Characteristics
DT	:	Decision trees
ID3	:	Itera-gaveDichotomise3
MMH	:	maximum marginal hyper plane
ARIMA	:	Auto Regressive Integrate d Moving Average
ANN	:	Artificial Neural Network
FP	:	False Positive
TN	:	True Negative
FN	:	False Negative
LOC	:	Line of Code
RF	:	Random Forest
NB	:	Naive Bayes
LR	:	Logistic Regression
LDA	:	Linear Discriminant Analysis
MLP	:	Multilayer Perception

DM	:	Data Mining
FL	:	Fuzzy Logic
NFL	:	No free Lunch Theorem
ELA	:	Ensemble Learning Algorithm
SBPRN	:	Stochastic Belief Poisson Regression Network
SBN	:	Stochastic Belief Networks
BM	:	Bayesian Model
ANN	:	Artificial Neural Network
PSR	:	Phase space reconstruction
MRMR	:	Minimal Redundancy Maximal Relevance
ANFIS	:	Adaptive neuro-fuzzy inference system

Chapter 1

Introduction

CHAPTER 1

INTRODUCTION

1.1 Rainfall Prediction

Rainfall is essential to the survival of all living things. It is important not only for humans, but also for animals, plants, and all other living things. Water is probably one of the most natural resources on the planet, and it plays an important part in agriculture and farming. Changes in climatic conditions, as well as rising greenhouse gas emissions, have made it harder for humans and the planet to get the appropriate quantity of rainfall to meet human requirements and continue to utilize in daily life. As a result, it has become critical to analyze shifting rainfall patterns and attempt to forecast rain not just for human needs but also for environmental purposes to forecast natural disasters that might be caused by unexpected heavy rains. [10].

1.2 Weather forecasting

Weather forecasting is the process of predicting the future state of the atmosphere at a specific location by applying the current technology and methods, applied to the atmosphere's present state such as humidity, temperature, and wind are gathered. Meteorology is the study of the earth's atmosphere and the distinctions in moisture and temperature patterns. In the process of meteorology, the data collected over the present atmospheric conditions are used to determine the weather forecasts. The major challenges with weather forecasts lie in the atmosphere's chaotic nature, and inadequate knowledge of the process and the forecast range to be predicted. In the case of Automatic weather stations or trained observers, initially, the process starts with observing the surface of the earth's atmosphere followed by collecting information like wind speed and direction, temperature, precipitation, and humidity. Varied numerical models or computer simulations are further applied to the data gathered to make the meteorological analysis. Computer simulations or calculations based on weather forecasting models start with gathering the present weather information and advance toward weather forecasting for the future by exploring atmospheric physics and fluid dynamics [117][7].

1.2.1　Different Weather Forecasting Methods

A. Synoptic Method: In this type of weather forecasting, the proper study of last weather forecasts from a wide area can be done. The current weather conditions are compared with the previously existing methods, here the weather predictions are made based on the current scenario and it behaves very much like the analogous situation in the previous.

B. Statistical Method: In this type of weather forecasting, regression equations relationships are made amid different weather elements and the upcoming climate conditions. Predictions are usually chosen based on a potential physical interaction with the predictands.

C. Numerical Weather Prediction Techniques: the motto of the numerical weather prediction technique is, to forecasts the weather by applying statistical models of the atmosphere as well as oceans based on the present weather conditions. The state of the atmosphere is represented in this system by a group of equations depending upon the physical laws of airflow, air pressure, and other related data. This method is best suitable for medium-term forecasts.

1.2.2 Weather Predictions on rainfall

Rainfall, stage a substantial character in the climate system as it directly influences other major factors such as agriculture, ecosystems, and water resource management. Often, if not properly managed, the heavy rainfall may also lead to natural disasters like floods, mudslides, landslides, and so on. Every year, the disasters caused by these heavy rainfalls are severely affecting the human-life and infrastructure [15][9].

For example, Wuhan is a large central city located at a longitude ranging from 113°41′ to115°05′ and a latitude ranging from 29°58′ to 31°22′. Wuhan generally possesses subtropical humid monsoon weather [59]. Wuhan was affected by El Nino in July 2016. Amid June 30th and July 6th, this massive rainfall reached 582.5mm, which is recorded to be the highest weekly rainfall of Wuhan to date [87]. 14 citizens died in this disaster and the economic loss that occurred was approximately 2.265 billion dollars. This instance proves that rainfall prediction is one of the major problems that need to be addressed to avoid or abate human as well as economic loss [15].

WMO (World Meteorological Organization), categorizes weather forecasting as Real-time which ranges from 0 – 2 hours of forecasting, Short-term forecasting which predicts the weather ranges between 2 – 72 hours, Middle-term forecasting which ranges from 72 – 240 hours and Long-term forecasting where prediction ranges from 10 to 30 days which further extended to 2 years. Many weather forecasting models provide the values of the current and the next day's weather [7]. The hourly-based weather forecasting models are failing in predicting accurate climatic information, especially prior warnings during a thunderstorm. It is vital to generate prior warnings from the weather forecasting models, which enable us to taking quick decisions regarding disaster management. In preventing and minimizing the loss of natural disasters like floods, more simple and accurate weather forecasting models are required [15].

1.2.3 Rainfall Influence factors

Rainfall is the major factor for water for farming activity all over the world. The three major features of rain that vary from dwelling to dwelling, year to year, periodically, and nocturnal today are rainfall amount, intensity, and frequency. Hence, knowledge of these three characteristics plays a vital role in rainfall prediction [118].

A lot of information including the day-to-day observations regarding rainfalls is available in various meteorological centers. In terms of monthly, and annually related to critical places, droughts, storms, and floods. Today, as the traditional approaches have been combined with emerging tools and technologies, it became easy to maintain even large volumes of data. Even though a large volume of data is maintained, but still there are several issues to address especially, effective prediction [118]

Meteorologists can't predict rainfall effectively by considering historical data such as the intensity of rains, frequency, and atmospheric conditions. In addition to the atmospheric conditions, the extent of effective or ineffective rainfall also depends on the field of agriculture, cropping patterns, soil types, and so on [118].

These difficulties are leading to unnecessary confusion in metrics, definitions, concepts, and the corresponding analysis. Hence, standard measurement techniques and the related nomenclature is required to recognize the significant rainfalls in a better manner and to transform the overall rainfall into ample rainfall to the maximum possible extent [118].

In general, normal rainfall occurs as the sun's energy heats the earth's surface, thereby, causing the water from varied water sources like lakes, ponds, rivers, and so on to evaporate and form into water vapor. As the land gets heated up with the sun's energy, the air above the land becomes warm, causing the air to expand and rise. Further, as the air rises high, it slowly cools and condenses. The condensed air forms cloud high in the atmosphere and this process is known as condensation. The water finally precipitates back to earth in the form of snow, fog, and rain. These rains are very common in tropical areas ranging amid the Tropic of Cancer and the Tropic of Capricorn. [119].

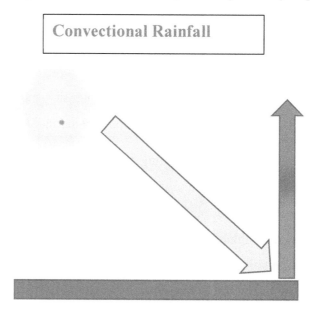

Fig1.1: Convectional Rainfall

Stage 1: The Sun heats the ground and as a result, the warm air rises.

Stage 2: The air thus raised, further cools and water vapor condenses forming the clouds.

Stage 3: As the condensation point is reached, the tiny water droplets convert Into large clouds.

Stage 4: The water droplets will grow heavy as the droplets condense with one another. As the heavy droplets are unable to sustain, they fall on earth as rainstorms.

Convectional rainfalls usually occur when the sun heats the ground. This is the main reason, why places like Amazon Rainforest, usually experience heavy rainfall in the afternoon [119,12].

1.3 Introduction to Machine Learning

Arthur Samuel (1959). Machine Learning is a field of study that gives computers the ability to learn without being explicitly programmed.

In recent times, machine learning has gained a lot of importance because of its capability in finding solutions to most conventional engineering problems. Machine learning algorithms train the computers on data and utilize statistical analysis to automatically predict output for the new input. This feature of machine learning simplifies the systems to build models for decision-making.

Generally, conventional engineering problems are domain-specific. So, domain-specific knowledge and optimized algorithms are required in building a statistical model. The statistical model, thus generated allows the machine to make automatic decisions for the new inputs and predict the output with good accuracy [94][17].

1.3.1 Categories of Machine Learning

A. Supervised machine learning.

The supervised learning models generally possess both raw input data as well as the corresponding results and are trained on a labeled dataset. The input Training and test

5

datasets are created first. The test dataset is used to test the model after it has been trained with the training dataset. That is, the trained model is employed on the test dataset treating it as new data. The results thus, obtained are compared with the actual output to assess the accuracy of the model. In supervised learning, the accuracy of the model increases with the perfection maintained in carrying out training and testing activities [115].

B. Unsupervised learning

Unsupervised algorithms are nothing; in this data, there are no labels for learning algorithms. These algorithms find hidden patterns depending on the feature's means. There are three types of cluster K-Means Clustering, Hierarchical Clustering, and Probabilistic Distance Clustering.

C. Reinforcement learning

It is a kind of trial-and-error method. It is taking some input and gives output. But the output is not certain and optimal. Based on the different actual and predicted values, further input will be supplied. The process will continue up to the derived output achieved.

D. Semi-supervised learning

Semi-supervised learning is a vital type that is in between the supervised and unsupervised machine learning. It works on the data that consists of few labels; it majorly consists of unlabeled data. As labels are costly, but for the corporate purpose, it may have few labels.

The basic disadvantage of supervised learning is that it requires hand-labeling by data scientists, and it also requires a high cost to process. Unsupervised learning also has a limited spectrum for its applications. To overcome these drawbacks, the concept of semi-supervised learning is introduced. In this algorithm, training data is a combination of both labeled and unlabeled data.

1.4 Machine Learning algorithms

Machine learning is the domain having many algorithms that they are supporting for machine learning applications. Here mention a few of the algorithms used in the thesis work.

K - NN algorithm

Support vector machine

Decision tree algorithm

Logistic regression

Random Forest algorithm

Bagging and boosting.

1.4.1 K-Nearest Neighbor (KNN)

It is self-styled learning. KNN is an easy and simple technique. It is used when the dataset is in a homogeneous type. It purely depends on the distance-based function and popularly uses the Euclidian distance function. The query point will be assigned to the nearest neighbor cluster based on its distance.

KNN is giving only local optima, but it fails to give the global optima value.

$$J = \sum_{i=1}^{m} \sum_{k=1}^{K} w_{ik} ||x^i - \mu_k||^2 \quad (1.1)$$

Equation (1.1)(1.2)& (1.3) If data point X is part of cluster k, w_{ik}=1; else, w_{ik}=0. In addition, k is the cluster centroid.

$$\frac{\partial J}{\partial w_{ik}} = \sum_{i=1}^{m} \sum_{k=1}^{K} ||x^i - \mu_k|| \quad (1.2)$$

$$w_{ik} = \begin{cases} 1, if \ k = argmin, ||x^i - \mu_k||^2 \\ 0, otherwise \end{cases} \quad (1.3)$$

1.4.2 Support Vector Machine

7

The objective of the SVM is to identify a hyperplane for multi-dimensional space, which uniquely predicts the data points.

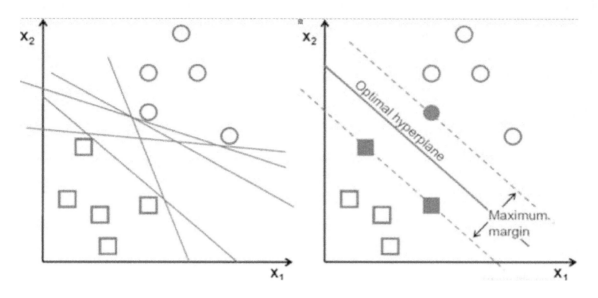

Fig 1.2: Support vector machine hyperplane.

To separate the data space, many feasible hyper lines could be chosen. aim to find a plane that has the maximum distance between data points of both classes. Maximizing the gap provides some added value so that future data values can be classified with more precision value.

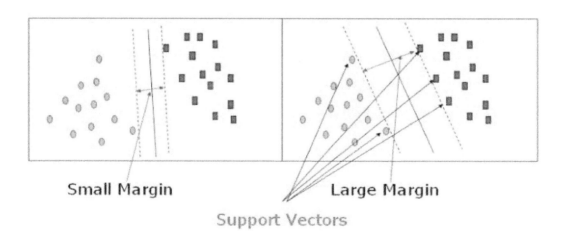

Fig 1.3: Support vectors

Decision lines are also called divisive planes that help to form groups. Data points on either side of the planes belong to different classes. The hyperplanes dimension depends on the number of features.

1.4.3 Decision Tree Classification

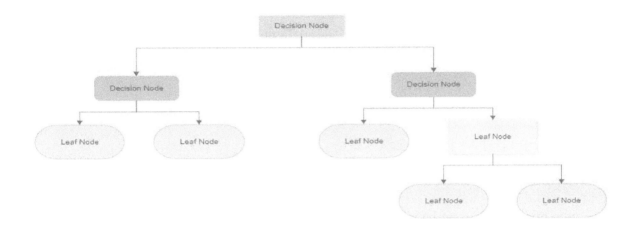

Fig1.4: Decision tree

A Decision Tree is a **guided technique, used both for classification and regression problems. But mostly decision tree is preferably used in classification. It is just like a tree-like structure where the internal nodes have the attributes of the data points, branches indicate the decision rules, and the leaf nodes or end nodes represent the class label values.**

The decision Tree is a graphical representation of the model. It contains two types of nodes. Internal nodes, having multiple branches depend on several features of the dataset. It is also called testing nodes or decision nodes. It is called a decision tree because it resembles a similar tree with the root node and expands based on decision criteria. The lowermost level of the decision tree contains leaf nodes having the decisions. To construct the decision trees, the following algorithms have been used.

a. CART
b. C 4.5
c. C 3.5
d. ID 3.0

Decision trees are easy to understand and implement. It is a robust technique for noisy and large data sets.

1.4.4 Logistic Regression (LR)

Logistic Regression is a simple numerical approach and is referred to as a standard statistical approach in computing credit scores (Lessmann et al., 2015). LR is majorly used in resolving regression problems and default as well as non-default binary classification problems. The motto of the LR model is to obtain the logarithm of the proportion of two desired probabilistic outcomes. Equation (1) is as follows:

$$\log[p(1-p)] = \beta_0 + \beta_1 X_1 + \beta_2 X_2 + \ldots + \beta_n X_n \qquad (1)$$

In which, log [p (1- p)] refers to the dependent variable and is defined as the logarithm of the ratio of two desired probability outcomes, with p representing the default probability. β_i is considered as the coefficient of the independent variables X_i (i = 1, ..., n). The goal of the LR model is to estimate the conditional probability for a definite sample set related to a class [12]

1.4.5 Random Forest Algorithm

It is one of the supervised learning techniques and a popular machine learning algorithm. We can use Random Forest Algorithm for both classification problems and regression problems too. It can work depending on the concept of ensemble learning, which means that it is a procedure of joining multiple classifier models to get the solution for complex problems and to improve the model performance.

It is a classifier built on a set of individual predictors. The data set is partitioned into parts. The predictors are developed on their partitions. The final classifier is the aggregation of all individual predictors, using majority voting and average in finding the query point class label or value.

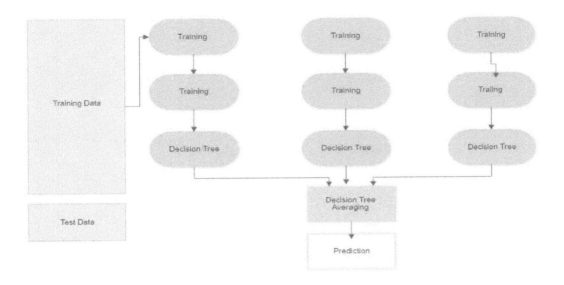

Fig 1.5. Random Forest Algorithm.

1.5 Ensemble uses two types of methods.

1.5.1 Bagging

It is an ensemble technique that generates a different subset for training from the whole dataset without replacement. It is using majority voting concept in its classification process. An example of bagging is the random forest algorithm, which uses the idea of bagging. Bagging stands for Bootstrap aggregation. Bootstrap is part of the dataset selected randomly by using an appropriate sampling technique. Each bootstrap will be used for a separate model. The final output of the model is the majority value of all individual models. i.e., It is aggregating the results of all models.

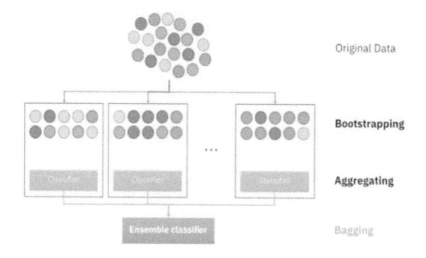

Original Data

Bootstrapping

Aggregating

Bagging

Ensemble classifier

Fig. 1.6 Bagging Technique

Bagging algorithm

It generates a combination of predictors for learning, in which every classifier has an equal weight in prediction.

Input:

- X, training samples.

- N, No. of classifiers in the ensemble.

- Base classifiers

Output: An ensemble model, E*.

Procedure:

1. Initiate k models.
2. Create a bootstrap sample, Xi, by sampling X with replacement;
3. Generate model E* with the usage of Xi;
4. End.

To apply the ensemble on sample x

Case 1: If it is categorical data for classification, the individual classifiers classify the given sample x and majority voting is used for final classification.

Case 2: If it is the prediction for numerical data sample x, then each predictor will give its value for x and the average value of all the predictors will be used.

1.5.2 Boosting

It is a sequential learner. It combines weak learners into strong ones by giving more weight to weak classifiers. The process of adding the learners till the final model achieves trusted accuracy. Examples are XG BOOST and CAT BOOST. Boosting means converting a weak one to a strong one. It decreases the error and increases the accuracy. Weak learners are identified, and their corresponding weights are incremented. In the process of training, Boosting gives more importance to less accurate models and vice versa.

1.6 Predictive Analytics

Predictive analytics utilizes historical data combined with varied machine learning and artificial intelligence algorithms to forecast future results in prior. A mathematical model is generated by feeding it with historical data to find hidden patterns and critical trends in data. The model is thus applied to current data to foresee future results [33].

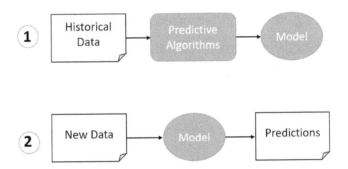

Fig 1.7: Predictive Analytics Models

The information generated from predictive analytics enables companies in enhancing their business applications. It helps in identifying the actions required for positive operational changes. The predictive analysis aids the analysts to foresee the consequences of a change. It assists the analysts in understanding whether the change made will increase the revenue and improve the operations or not [32]. A sincere and upfront work setup is

13

required in building accurate and worthwhile predictive analytics. Predictive analysis also involves a team of people who understands the business problem, the data that is required to carry out analysis, and the models to be developed and sophisticated. The team plays a vital role in converting predictions into an act for better outcomes [119].

1.6.1 Predictive Analytics Techniques

The risks linked with opportunities can be assessed by analyzing the patterns of predictive models in transactional as well as historical data. These predictive analysis techniques are utilized to find the hidden trends in the data. Enhanced predictive models are developed by integrating multiple predictors that can be used to predict future probabilities with great reliability. By combining the statistical techniques and business knowledge with the perceptions formed from the predictive analysis, a healthcare organization can successfully estimate the cost of a product and insurance in prior [35].

Data profiling and transformations, Time series tracking, and Sequential pattern analysis are the major techniques involved in Predictive analytics. The technique, Data profiling, and transformations include the functions that are responsible for altering the row and column values, exploring the pattern dependencies, combining fields, formatting data, and aggregating the records in forming rows and columns. Sequential pattern analysis identifies how the rows of data are related. This approach governs the serial items that repeatedly occur throughout all ordered transactions with time. A time series track can be defined as a series of values specified at periodic time intervals concerning a particular distance [104]. Classification Regression is also a good example of a Standard Predictive analytic approach.

Classification technique in data mining assigns the data from a collection to a target class or category. The motto of the classification approach is to predict the target class of every item present in the given collection with good accuracy. Training and testing using data with established class labels are two of the three processes involved in classification. The training data uses a subset of the data to build a model. Following that, testing data will predict the class labels and accuracy of the model, which was constructed using training data. The model is satisfactory throughout the deployment phase. The model is then tested with unlabeled data. For the building of classification models, a popular classification method is used. [104].

Regression is a statistical approach used to model the relationship between one or more independent and dependent variables with continuous values. Multiple Linear Regression is a complex approach that employs a variety of input variables to deploy more complicated models that are higher-order polynomial equations. [104].

1.6.2 Applications of Predictive Analytics

Predictive analytics is used to predict future trends based on data behavior. It is the confluence of Statistics, AI, and Machine learning disciplines. It is popularly used in decision-making by business organizations and entrepreneurs. It is a kind of advanced data analysis. Nowadays this is very much required in the present day competitional business environment. Risks are detected and possibilities for future demands are explored through the use of historical data and transactional data trends. Businesses can successfully understand the data at their disposal for their profit by using predictive analysis. [12]. The process involved in Predictive Analytics is shown in fig 1.8

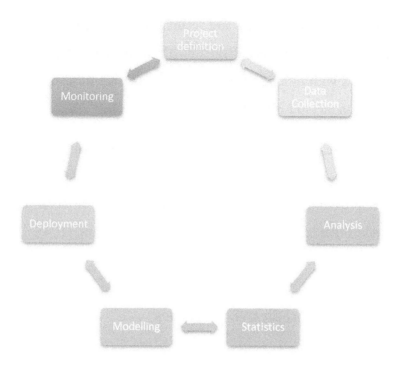

Fig1.8: Process of Predictive Model

Predictive analytics applications are.

1. CRM: Predictive analytics helps most companies in meeting their objectives such as sales enhancement, marketing campaigns, and customer services. The application of predictive analysis eased the management of customer relationships. Throughout the customer life cycle, predictive analytics is applied at different phases such as acquiring relationships, relationship development, retaining, and customer win-back [119].

2. Health care: In recent times, the use of predictive analytics in health care is drastically increasing. This technique is not only used to determine the complication but also, helps in preventing and reducing the probability risk. Predictive analytics aiding medical experts in treating the most life-threatening diseases such as asthma, diabetes, and so on. Better and more accurate clinical decisions are easily made by applying predictive analytics in health care.

3. Collection Analytics: In the collection analytics domain, predictive analytics is majorly used in finding the collection agencies, varied contact approaches to reach them, and legal activities required to improve recovery and minimize the collection cost.

4. Cross-Sell: Predictive analytics helps in obtaining consumer behavior. These approaches present a detailed analysis of a consumer such as the frequent items they purchase, expenditures, and products of interest. Further, consumer analysis thus made helps in making cross-sales or in selling added products to consumers. Hence, this approach supports the organizations in making ultimate profits.

5. Fraud Detection: Financial and Insurance agencies are applying predictive analytics in detecting online as well as offline frauds, finding inaccurate credit applications, identifying untrue insurance claims, and finding thefts. Most financial organizations are applying predictive analytics to handle security concerns and to burnish day-to-day operations.

6. Risk Management: The main objective of predictive analytics in Risk management is to predict the ROI (Return on Investment). So, the approaches help in assessing the risk involved as well as the maximum returns that can be earned on an investment.

7. Direct Marketing: Continuously changing marketing trends is the biggest challenge for any business organization. Predictive Analytics assists organizations in finding varied product versions; marketing trends, different communication methods and the right time to target a customer, and so on. So, these techniques help organizations not only to run their business successfully but also to earn great profits.

8. Underwriting: Predictive analytics simplified the consumer acquisition process by predicting the upcoming risk the consumer may face. Predictive analytics is applied to consumer application data to predict future risks such as the chances of illness, economic failure, and default of loan/insurance [119].

1.7 ORGANIZATION OF THE THESIS

The thesis is structured into seven chapters. The chapters of the thesis are outlined as follows:

Chapter 1 Introduction, this chapter discussed the basic problem domain nomenclature and fundamentals of machine learning algorithms.

Chapter 2 Literature Review deals with the review of various contributions to rainfall prediction techniques and data cleaning processes.

Chapter 3 System analysis and system Design deals with the problem statement, objectives, proposed methodology, and the overall system architecture for rainfall prediction.

Chapter 4 Data preprocessing and feature enhancement deals with Data Preprocessing and Feature Extraction Techniques.

Chapter 5 Ensemble machine learning models focus on Time-series data analysis by using ensemble machine learning models.

Chapter 6 A hybrid firefly optimization with MLP and VAE mechanism deals with accurate rainfall prediction.

Chapter 7 of the thesis on conclusion and future work.

1.8 Conclusion

The various concepts consist of an introduction to weather forecasting, various weather forecasting methods, rainfall influence factors, an introduction to machine learning, and various machine learning algorithms which are used for prediction, ensemble models, and predictive analytics techniques, and their applications have been included in the introduction chapter. The next chapter will discuss various existing prediction methods and techniques used for rainfall predictions and their merits and demerits are identified.

Chapter 2

Literature Review

CHAPTER 2

LITERATURE REVIEW

2.1 Introduction

This chapter presents the study and survey done on Machine Learning books, Rainfall prediction scholarly articles, and a critical evaluation of the works concerned with the research problem being investigated. Literature reviews comprised in this chapter provide an overview of sources that are explored while researching rainfall prediction models.

2.2 Survey on Rainfall Prediction

Literature reviews provide an overview of available sources we have surveyed while searching a specific topic on different rainfall prediction models and demonstrate research gaps in a weather prediction study.

A Literature Survey has been carried out by referring:

1. One hundred and three (103) Journals and Conference papers were collected and studied from IEEE, SCIENCE Direct, Springer, and UGC Approved Journals and other sources.

2. By attending five Faculty Development Programs on Machine Learning Algorithms using Python.

Hewage et al.[80][35] demonstrated how different machine learning techniques succeeded in predicting rainfalls. In this paper, popular machine learning algorithms like Auto-Regressive Integrated Moving Average (ARIMA), Self- Organizing Maps, Support Vector Machine, Logistic Regression, and Artificial Neural Networks are discussed. The experimental results proved that ARIMA and Artificial Neural Network (ANN) models successfully predicted the rainfalls by using the techniques such as Cascade NN and Back Propagation NN.

Kader et al.[2] implemented a novel, flood detection, and alerting system by using IoT techniques. With the advancements in emerging technologies like machine learning (ML), it became very easy to differentiate the normal as well as abnormal behavioral characteristics of a system. The author discussed various surveys and studies carried out on flood issues. The paper also presented the techniques such as Neural Networks, which are proven to be efficient in predicting rainfalls.

Hatem et al.[3]presented a hybrid technique, that combines both Multi-Layer Perceptron (MLP) and Particle Swarm Optimization (PSO) to make rainfall prediction. It is proved that the proposed technique, succeeded in enhancing the performance of the network along with rainfall prediction. The proposed technique is compared with the BP (Back Propagation) algorithm like LM (Levenberg-Marquardt) and the experimental results demonstrated that RMSE for MLP-based PSO is 0.14 whereas RMSE for MLP-based LM is 0.18.

Ricardo et al.[5] demonstrated the comparative analysis made on monthly rainfall predictions in islands using varied machine learning algorithms. The model used two predictors, global predictors like North Atlantic Oscillation Index (NAO) and local predictors like local Geopotential Height (GPH). The model is tested on the Canary Islands (The island of Tenerife) by considering the dataset collected over four decades. The model is tested using predictive models such as Extreme Gradient Boosting and Random Forest. The results are analyzed in terms of interpretability, accuracy, and kappa and the results proved that local predictors have a high influence over global predictors in rainfall prediction.

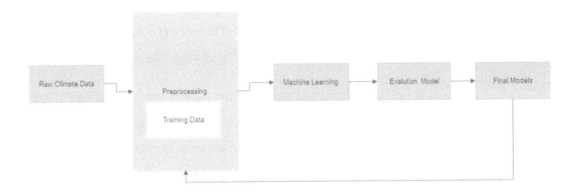

Fig.2.1: General flow chart of the proposed method

Ahmadreza et al.[47] proposed PV-RNN, a new variation RNN model stimulated by predictive-coding ideas. The proposed model is capable of extracting the unseen probabilistic structures from changing temporary patterns by altering the latent states dynamically. The main motto of the architecture is to address the issues with Bayes RNNs. Firstly, to make meaningful patterns from the hidden variables and secondly, to

20

derive future observations from these hidden variables. The proposed model is trained using adaptive vector mirroring and tested using the error regression technique. The model is tested on two datasets and the results showed that the model performed as a random approach for low values whereas the network outperformed for intermediate values.

Cramer et al.[20] has discussed how machine learning-based intelligent systems can be employed in rainfall prediction. In this work, six standard machine learning algorithms, Genetic Programming, M5 Model trees, M5 Rules, Support Vector Regression, k-Nearest Neighbors, and Radial Basis are compared by applying them to datasets comprising rainfall time series collected from 42 cities. The experimental results proved that the machine learning-based intelligent systems outperformed in predicting rainfalls accurately and in detecting the hidden correlations amid different climatic conditions.

Yajnaseni et al.[23] presented the varied ways to forecast ISMR (Indian summer monsoon rainfall). The author discussed three different algorithms namely, SLFN(Single layer feed-forward neural network), RVFL(Random vector functional link network), and ROS-RVFL (Regularized online sequential network – RVFL). ISMR for the coming year is been predicted by applying all three algorithms SLFN, RVFL, and ROS-RVFL by considering the previous data. The same process is repeated on six different datasets comprising previous data of varied lengths ranging from 5 to 10 years. The experimental results have shown that, over SLFN and RVFL, ROS-RVFL generated better results in terms of efficiency and accuracy especially when the model is trained by considering 8 to 9 previous years' data.

Javier et al.[29] considered and estimated the performance of 8 different statistical and machine learning techniques in predicting the long-term daily rainfall in Spain possessing a semi-arid climate. The rainfall data is reconstructed for 36 years by considering each of the 17 different gauges independently. Now, the generated data is compared with actual observations, to identify the best hyper-parameters. The performance of the algorithms is assessed by taking different statistics and metrics connected with rainfall intensity and occurrence at different time ranges like daily, weekly, monthly as well as annual aggregation scales. The demonstrated results proved that the performance of the machine learning models is mostly depending on the hyper-parameters selected.

Sani et al.[86][36] demonstrated the comparative analysis of various machine learning algorithms. SVM, NB, RF, NN, and DT are applied to the rainfall dataset collected from Malaysia to forecast rainfall for different time ranges daily and monthly. The performance thus generated is compared and the experimental results proved that Neural Network (NN) is yielding better results compared to other algorithms in terms of accuracy.

Using the decision tree method, Geetha et al.[31] developed a unique weather prediction model. Predicting the weather is difficult since it is dependent on fluctuating atmospheric elements such as humidity, temperature, and wind speed, rainfall that changes from time to time. Hence, in this paper, a new model using a decision tree is introduced to predict weather phenomena such as rainfall, fog, thunderstorms, and cyclones. The model enhances the chances of taking wise and quick decisions during natural disasters such as floods etc. The proposed model experiments with the Rapid Miner tool.

Martin et al.[37] introduced a new technique in which NWP, the numerical weather prediction models utilized machine learning techniques to forecast solar energy production. Initially, the technique is used to verify to what extent the prediction accuracy depends on the number of input NWP grid nodes. Further, the technique is used to predict solar energy production in an area where previous data is not available. The technique is a technique using different feature selection and machine learning algorithms and the results are evaluated.

The models are proven to predict energy production for new as well as existing stations, denoted by longitude and latitude.

Namitha et al.[68][39] proposed a new technique to handle huge volumes of weather-related data by utilizing Hadoop. The approach employed Artificial Neural Network on a Map-reduce framework to assess short-term rainfall forecasting. The technique uses the temperature and rainfall data of the current day and predicts the rainfall of the following day in prior. The rainfall and temperature data of India over the past 63 years from 1951- 2013 is used to carry out the study.

Nikam et al.[71] developed the data-intensive model utilizing data mining techniques. Unlike the general weather forecasting systems that require supercomputing power to predict weather conditions as well as rainfall, the author introduced a data-intensive model that predicts rainfall with less or moderate resources. Using the Bayesian method, tested on the dataset collected from IMD (Indian Meteorological Department), Pune comprising 36 varied attributes. The demonstrated results proved that only 7 attributes among 36 are more effective and the technique succeeded in predicting rainfalls with good accuracy.

Aftab et al.[4] presented how different data mining techniques are employed in predicting rainfall. The dataset is extracted from a weather forecasting website. The above-mentioned algorithms are applied to the dataset to predict rainfalls in Lahore city. The experimental results are assessed concerning recall, f-measure, and precision by considering different sizes of training and test datasets.

Table 2.1: MLP performance metrics.

Proportion	Class	Precision	Recall	F-Measure
10-90	No Rain	0.927	0.993	0.959
	Rain	0.709	0.173	0.278
20-80	No Rain	0.935	0.983	0.959
	Rain	0.61	0.28	0.384
30-70	No Rain	0.931	0.99	0.96
	Rain	0.681	0.221	0.334
40-60	No Rain	0.937	0.979	0.958
	Rain	0.579	0.303	0.397
50-50	No Rain	0.931	0.992	0.96
	Rain	0.724	0.22	0.337
60-40	No Rain	0.937	0.98	0.958
	Rain	0.587	0.304	0.4
70-30	No Rain	0.93	0.99	0.959
	Rain	0.67	0.212	0.322
80-20	No Rain	0.94	0.984	0.962
	Rain	0.619	0.296	0.401
90-10	No Rain	0.935	0.997	0.965
	Rain	0.846	0.21	0.336

Kasun et al.[83][40] presented the survey and varied studies carried out on predicting energy production from renewable sources. Power production majorly depends on environmental parameters and hence, it is not possible to control it completely. In this paper, the author proposed various mechanisms and demonstrated how machine learning algorithms can be employed to predict power production. The machine learning algorithms can be successfully applied to address many challenges the power plants currently face like determining the best location, grid sizes, configuration, and so on. The author presented the survey on how varied machine learning methods addressed various challenges related to renewable energy production.

Parmar et al.[1] presented the comparative study and work of different rainfall prediction systems proposed by different researchers. Because of the changing nature of the atmosphere, statistical techniques are failing to forecast rainfall accurately. Hence, varied artificial neural network techniques replaced traditional statistical techniques in predicting rainfalls. The motto of this paper is to help even the non-experts in getting knowledge about different approaches and techniques that are implemented in the area of rainfall prediction.

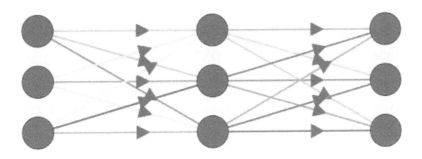

Fig 2.2: BPNN architecture with one hidden layer.

ManGalih et al.[85] enlightened the usage of varied deep-learning techniques in weather forecasting. CN (Convolution Network), CRBM (Conditional Restricted Boltzmann Machine), and RNN (Recurrence Neural Network) models are applied to the dataset collected from Indonesian Agency for Meteorology, Climatology, and Geophysics (BMKG) and National Weather Service Centre for Environmental Prediction Climate (NOAA) from 1973 to 2009. The experimental results of all the models are compared using the Fresenius norm concerning accuracy.

Jae-Hyun et al.[90][41] introduced a technique to forecast heavy rainfall in South Korea prior by 1- 6 six hours. This approach reformed the AWS data of the four recent years by normalizing them to numeric values amid 0 and 1. 0 represents no- heavy-rain whereas 1 represents heavy rain. The dataset collected for the years 2007- 2010 is divided into three parts. The data ranged from 2007-2008 used for training and 2009-2010 for testing.

Evolutionary algorithms like variant K-NN (K-VNN), K-Nearest Neighbor algorithm (K-NN), and support vector machine (SVM) are used to form the rainfall predictions. The results showed that SVM with a polynomial kernel generated accurate results over other algorithms.

24

By integrating Artificial Neural Networks with the right use of Genetic Algorithms, Solanki et al.[97] created a hybrid system. The evolutionary algorithm is used to efficiently train the neural network model and construct a connection topology between the input and output layers. The MLP (Multilayer Perceptron) method is used to perform prediction tasks, while the BP (Back Propagation) approach is utilized to train the model. The outcomes showed that the suggested method was successful in predicting rainfall with high accuracy.

Wei et al.[106] implemented a Hadoop Spark distribution framework by using big-data technology, to enhance the computation of typhoon rainfall prediction models. MLR (Multiple Linear Regression) and DNN (deep neural networks) are used to predict typhoon rainfalls. These models are combined with the Hadoop Spark distribution framework to handle big data and improve the computation speed. The model is trained considering the dataset extracted from 271 typhoon events over the period 1961 to 2017. The results showed that the proposed framework succeeded in predicting typhoon rainfalls with good accuracy and in enhancing the computation speed.

Lily et al.[21] introduced a rainfall classification and prediction model by using the three most popular machine algorithms namely SVM (Support Vector Machine), DT (Decision Tree), and ANN (Artificial Neural Network). The technique helps in finding the relationship between weather conditions and rainfall occurrences and predicting rainfalls. The model is tested on datasets extracted from the Chalermprakiat Royal Rain-Making Research Center from 2004 to 2006. The experimental results showed that C4.5, ANN, and SVM achieved 62.57%, 68.15%, and 69.10% accuracy correspondingly. The model also succeeded in classifying the rainfall into three categories as no-rain (0-0.1 mm.), few-rain (0.1- 10 mm.), and moderate-rain (>10 mm.)

Rani et al.[80] proposed a novel methodology to reduce the computation time required for rainfall forecasting. The technique used a two-level clustering technique. The first step is to use SOM (Self Organized Maps) to develop a prototype and the

second step is to divide the large dataset into clusters. The model used ID3 in applying clustering on SVM and SOM and the results generated are analyzed and it is proved the technique successfully reduced the computation time required in predicting rainfalls compared with direct clustering.

Neelam et al.[61] presented a comprehensive study performed on how the time series data can be used effectively in rainfall prediction. Large volumes of time series data are available in various meteorological stations. This paper demonstrated varied statistical and data mining techniques that can be applied to the available time series data to forecast rainfalls accurately and also discussed how this time series data can be converted into information for future algorithms.

Aswin et al.[11] developed a rainfall prediction model using deep learning architectures. The model used the architectures Convolution Network and LSTM. The model uses the available time series data and past events to predict future events in prior. The model is tested using the Convolution Network and LSTM deep learning techniques and the results showed that RMSE for Convolution Network is 2.44, whereas RMSE for LSTM is 2.55.

Suhaila et al.[113] demonstrated how the efficiency of the various machine learning algorithms changes with the metrics such as the size of the training and test sets, and the period of Prediction like monthly, weekly, and so on. In this paper, the most common machine learning algorithms like Decision Trees, Naïve Bayes, Random Forest, Neural Networks, and Support Vector Machines are analyzed under different metrics. The algorithms are tested on the dataset extracted from multiple meteorological stations located in Selangor, Malaysia. The demonstrated results showed that Random Forest outperformed all when trained using a small training dataset (10%). From 1581 records, it predicted 1043 instances correctly.

Cramer et al.[20] implemented a new rainfall prediction algorithm namely, Decomposition Genetic Programming (DGP). The algorithm initially, decomposes the rainfall problem into sub-problems. DGP enables to focus on each sub-problem individually, thereby, reducing the complexities with large volumes of data.

The algorithm is rigorously tested on data collected from 42 cities in the USA and Europe. The results demonstrated that DGP is generating the best results compared to other algorithms in terms of accuracy and efficiency.

Kavitha et al.[27] presented the comparative study of data-driven modeling approaches like rainfall time series modeling (SARIMA) and multilayer perceptron neural networks (MLP-NN) in forecasting rainfalls. The models are analyzed and compared by considering various statistical parameters such as Mean Absolute Error MAE, Root Mean Square Error RMSE, BIAS, and Coefficient Of Correlation CC.

Yajnaseni et al.[102] demonstrated the use of artificial intelligence approaches in

26

rainfall prediction. Artificial Neural Networks (ANN), Extreme Learning Machines (ELM), and K-nearest Neighbor (KNN) are used to predict both monsoon (June-September) and post-monsoon (October-December) rainfalls. The models are tested on the dataset extracted from Kerala, India over the period 2011 – 2016. The experimental results proved that the ELM approach is yielding better results compared to ANN and KNN with minimal mean absolute percentage error. In the case of summer monsoon rainfall the error score is 3.075 whereas, for post-monsoon, it is 3.149 respectively.

Chauhan et al.[16] presented the study of Data Mining Techniques in Weather forecasting and the varied advantages of using them. The paper offered a survey of the various techniques and approaches different researchers employed in weather forecasting. All the available techniques and algorithms are compared and the author concluded that k- mean clustering and decision tree proved to be the best concerning higher prediction accuracy compared to other techniques of data mining.

Sethi et al.[91][48] presented the comparative study of different machine learning algorithms in rainfall prediction. Agriculture plays a very vital role in the Indian economy. Hence, early Prediction of rainfalls will surely enable farmers in making several decisions like which crop to harvest, required water resources, and so on. ANN, Regression analysis, and clustering algorithms gained a lot of popularity in weather as well as rainfall prediction. The author proposed MLR (Multiple Linear Regression) models for rainfall prediction in prior.

Dong et al.[26] proposed a new algorithmic solution to weather forecasting. The methodology is a hybrid technique that uses multiple machine-learning algorithms to carry out varied tasks. As an initial step, PSR (Phase space reconstruction) is used to convert the available time series data into a set of numerical variables. In the second step, IVS (Input variable selection) and MRMR (Minimal Redundancy Maximal Relevance) are used to automatically identify the optimal variables that are having a high

influence on weather forecasting. Lastly, K-means clustering is used to train ANFIS (Adaptive neuro-fuzzy inference system) and is used to make rainfall predictions. Further, PSO (Particle swarm optimization) algorithm is used to enhance the performance of the model. The results thus generated proved that the model successfully made rainfall prediction accurately and efficiently.

Table 2.2: Main setting parameters for training.

Main Techniques	Parameters	Settings

Hyperparameter of PSO	Intertia weight, Intertia weight damping ratio, Global Learning Coffeicient	W0=1 Wdamp=0.99 C1=1
Hyperparameters of ANFIS	Number of fuzzy sets in each input variable	Nfuzzy=5
Hyperparameters of K-Means	Number of clustering centers	K=6

Alexis et al.[28][49] presented a novel mechanism to analyze solar irradiation. Eleven different machine learning and numerical tools are individually employed from 1 to 6 h to predict solar irradiation on an hourly basis. To generalize the conclusions and to make them more accurate, the solar data is extracted from three different sites namely Ajaccio with low, Tilos with medium, and Odeillo with high meteorological parameters. The models are analyzed and related concerning skill score, Mean absolute error, and normalized root means square error. The experimental results showed that multi-layer perceptron and auto-regressive moving average proved to be best under low variability, bagged regression tree, and auto-regressive moving average outperformed for medium variability whereas, for high variability, random forest and bagged regression tree are proved to be most efficient.

SueEllen et al.[34] presented the first automated weather forecasting system, DICast (Dynamic Integrated Forecasting) developed at NCAR (National Center for Atmospheric Research). NCAR is one of the best research centers which addressed many of the weather forecasting challenges by employing machine learning algorithms. Today, several companies are using DICast in many of their applications. In addition to DICast, NCAR also developed a set of artificial intelligence technologies such as wildland fire forecasting, surface transportation, and renewable energy.

Sundaravalli et al.[98][50] implemented an innovative technique to predict crop production by predicting the rainfall in prior. In this paper, K- Means clustering algorithm, fuzzy classification algorithm, and a hybrid technique, Neuro-Fuzzy merged with the genetic algorithm are discussed. The three models are analyzed and tested. The demonstrated results showed that all three algorithms thrived in predicting crop production by predicting the rainfalls prior and Neuro Fuzzy proved to be the best in terms of accuracy compared to K-Means and fuzzy algorithms.

Sachindra et al.[83] introduced a novel statistical model for data downscaling. The model was applied to the data collected from 48 stations located in different regions of the Australian State of Victoria. The stations extracted the data from the regions belonging to different climatic conditions such as dry, intermediate, and wet. The downscaling models are trained considering the data collected over the period 1950-1991 and are tested on the data over the period 1992-2014. The following four machine learning algorithms namely, ANN (Artificial Neural Networks), GP (Genetic Programming), RVM (Relevance Vector Machine), and SVM (Support Vector Machine) are used to implement the downscaling models. Based on the demonstrated results, it is shown that RVM is more efficient in implementing downscaling models over SVM, ANN, and GP.

Shah et al.[92] discussed different machine learning algorithms and modes implemented for precipitation prediction. The main motto of meteorological research is weather forecasting depending on various parameters like temperature, humidity, precipitation, and so on. A varied set of meteorological parameters are required to predict rainfall using machine learning algorithms. In this paper, different machine learning techniques are analyzed and showed that NN (Neural Networks) and ARIMA are best in making accurate precipitation predictions over a period ranging from daily to weekly basis. Whereas, Random Forest suits best if the Prediction is to be made for the next season prior.

Table 2.3: Performance comparison between various algorithms.

Parameter	ANN	DTN	Clustering	GA
Performance	Maximum	High	Moderate	High
Supervised Learning	Yes	Yes	No	Yes
Unsupervised Learning	Yes	No	Yes	Yes
Computation Speed	Fast	Fast	Slow	Fast
Cost-effectiveness	Yes	Yes	Yes	Yes
Accuracy	More	Less	More	Moderate
Complexity Level	Complex	Less	Moderate	Complex
Fault Tolerance	Yes	Yes	Yes	Yes

Nature	Analytical	Analytical	Descriptive	Analytical
Domain Area	Complex	Simple	Complex	Complex

Tharun et al.[24][52] implemented a novel technique to assess rainfall intensity by utilizing regression techniques and statistical modeling. The model is tested on the dataset extracted from Coonoor, Tamil Nadu. The regression techniques, DT (Decision Tree), RF (Random Forest), and Support Vector Regression (SVR) are used to build the prediction model. The model is trained on the dataset comprising varied parameters such as cloud speed, wind direction, wind speed, humidity, and temperature that are extracted from day-to-day observations. The prediction model is implemented in the python platform and is analyzed by considering Adjusted R-square and R-square values. The forecasting model implemented using the RF regression technique produces deficient results compared to DT and SVR models.

Xiang et al.[110] developed an innovative rainfall prediction model by combining machine learning algorithms with EEMD (Ensemble Empirical Mode Decomposition). In this approach, initially, the available time-series data is assessed from short-period to long-period time spans using EEMD. The datasets are extracted from different meteorological stations situated in China. The model is tested by employing different machine learning algorithms on the data collected using EEMD. The experimental results proved that SVM (Support Vector Machine) and ANN (Artificial Neural Network) produced accurate results concerning short-period and long- period respectively.

Nabila et al.[114] implemented a rainfall prediction model by employing Naïve Bayes, Artificial Neural Networks, and Predictive Decision Tree models. The model is tested using the dataset extracted from a meteorological station located in Subang over the period from January 2009 to December 2016. The dataset possesses 2922 observations including five major factors like minimum and maximum temperatures, evaporation, cloud, and wind speed. The results of the three algorithms are related and proved that the performance of the predictive Decision Tree model is better when compared to other models with RMSE is 0.35 and a misclassification rate is 0.15.

Kodimalar et al.[72][54] proposed a new technique to predict crop yield by predicting rainfall. The author demonstrated how machine learning algorithms can be successfully employed in rainfall and weather prediction. Further, the predictions made will enable

30

the farmers in deciding which crop to be grown for the coming season in prior and also predict the crop yield.

Radhika et al.[79] presented the application of SVM (Support Vector Machines) in weather forecasting. In this approach, the current day maximum temperature at allocation is explored to assess the maximum temperature of the coming day prior. The model is trained using a Non-linear regression model and by utilizing the kernel functions' optimal values, the model is tested for 2 to 10 days. The generated outputs are compared with MLP (Multi-Layer Perceptron) trained using a back-propagation algorithm and it is shown that SVM produced accurate results.

Ibrahim et al.[30] demonstrated the study done to verify whether it is feasible to apply machine learning algorithms on NCDC (National Climatic Data Centre) to predict weather conditions or not. The results were further compared with the traditional meteorological models. A varied set of popular machine learning techniques are employed to produce vigorous weather forecasting models for long periods. The models are tested on datasets comprising multiple meteorological parameters and the experimental results proved that AdaBoost, XG Boost, and decision tree produced better results concerning classification accuracy whereas the linear regression model proved to be best concerning regression task.

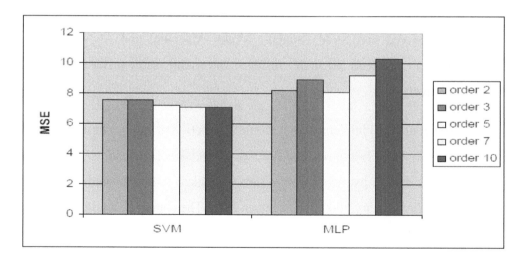

Fig. 2.3: Comparison between MLP and SVM for different orders.

Dongjin et al.[18] proposed a new air temperature forecasting model. To overcome the issues with NWP (Numerical Weather Prediction) such as correcting LDAPS, a local NWP model built in Korea, the proposed model uses support vector regression (SVR), random forest (RF), multi-model ensemble (MME) and artificial neural network (ANN) in building the forecasting model. The demonstrated results showed that the MME model had better generalization performance when compared to other machine learning algorithms.

ANANDHARAJAN et al.[9][55] has implemented an Intelligent Weather Prediction tool. The tool uses the parameters such as minimum and maximum temperatures, and rainfall for a sample period and the data is explored. The tool utilizes the machine learning techniques like linear regression to forecast the next day's weather with good accuracy. Recent studies showed that machine learning algorithms are producing more accurate results compared To traditional statistical approaches. Some of the models achieved 90% accuracy in predicting rainfalls.

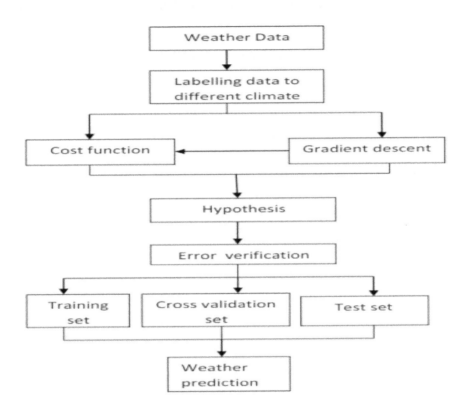

Fig.2.4: Classification process.

Ricardo et al.[8] introduced an innovative technique, to forecast solar energy by employing NWP (Numerical Weather Prediction) models as shown in fig 2.4. The proposed technique is explored using both the datasets extracted from Kaggle as well as the State of Oklahoma. The model is evaluated using Gradient Boosted Regression and Support Vector Machines models. The results proved that NWP models can even be applied to solar energy prediction.

Mark et al.[38] implemented two models, the linear regression model and the functional regression model to assess the maximum and minimum temperatures seven days prior.

Both models were outperformed when compared to traditional weather forecasting mechanisms. When the models are trained considering two days of data, the linear regression model produced accurate results. But, when the models are trained for four to five days, the functional regression model outperforms the linear regression model.

Wanie et al.[48][57] The repository's water level is generally constrained by precipitation. Environmental change, then again, may make flood or dryness in the repository attributable to the capricious amount of precipitation. To estimate precipitation information in TasikKenyir, Terengganu, many models, and approaches were utilized in this review. The review thought about two methodologies for estimating precipitation and created and looked at numerous Machine Learning (ML) models, just as investigating various circumstances and time ranges. Utilizing the Thiessen polygon to gauge the station region and anticipated precipitation, this review takes the normal precipitation from ten stations all through the review locale. The anticipating model depends on the accompanying suppositions: Bayesian Linear Regression (BLR), Boosted Decision Tree Regression (BDTR), Decision Forest Regression (DFR), and Neural Network Regression are four elective AI techniques utilized in the gauging model (NNR).

The precipitation, then again, was anticipated throughout an assortment of time skylines utilizing two separate AI calculations: strategy 1 (M1): Forecasting Rainfall Using Autocorrelation Function (ACF), and technique 2 (M2): Forecasting Rainfall Using Projected Error. Since it has the most elevated coefficient of assurance, R2, in the wake of setting the hyperparameter, BDTR is the best relapse intended for ACF in M1. Every day (0.9739693), week-by-week (0.989461), ten days (0.9894429), and month-to-month coefficients are all in the scope of 0.5 to 0.9. (0.9998085). Except for 10 days, where BDTR and DFR are more OK than NNR and BLR, in general, model execution in M2 shows that standardization utilizing Log Normal favored offers a respectable result in every classification. Therefore, it very well may be surmised that two separate strategies were utilized with different conditions and worldly skylines, with M1 showing more noteworthy precision than M2 while utilizing BDTR-demonstrating.

2.3 Some Prominent papers

Table 2.4: Literature survey summarization.

Method for the Study	Application Area	Journal Name	Publisher	Year	Compared Method	Problem Type	Performance Factor
BLR, BDTR, DFR	Forecasting rainfall using auto co-relation function	*Atmosphere (Basel).*	MDPI	2020	NNR	Forecasting rainfall using projected error	MSC
Random Forest and Gradient Boosting	Feature Extraction	*Appl. Sci*	SCI	2019	Random forest, ADA boost, gradient Boost	Feature enhancement	Time Complexity, Feature classification
Neural networks and ARIMA	Forecasting depends on various parameters	*Int. Conf. Parallel, Distrib. Grid Comput.*	Elsevier	2018	Random forest	Predictions for next season	Daily to weekly basis accuracies performance
DGP	Sub problem individually reduces the complexities with a large volume of data	*Appl. Soft Comput. J*	Elsevier	2017	SVM, LSTM	Decomposes the rainfall problem into sub-problems	Accuracy and efficiency
Classification and Prediction	high volatility and chaotic patterns that do not exist in other time-series	*Expert Syst. Appl*	Elsevier	2017	Support Vector Regression, Radial Basis Neural Networks, M5 Rules,	Classification and time series	Accuracy and Confusion Matrix

					M5 Model		
Random forest	The period of prediction like weekly, monthly so on.	*Int. J. Adv. Sci. Eng. Inf. Technol.*	search	2016	Decision tree naïve bayes	Classific ation problem	Precision and recall
predict weather phenomena	model using a decision making and Prediction	*IEEE Int. Conf. Comput. Intell. Comput.*	IEEE	2014	decision tree,ID3,R andom forest	Weather and Atmosp here	Accuracy and Confusion Matrix
Data Cleaning	data pre-sevieprocess ing and data transformati on	*Proc. Int. Conf. Comput. Intell. Model. Simul*	IC-EPSMSO	2013	SVM ,Baysian	Weather Data Pre Processi ng and Cleanin g	Fitness of data, Noise clearness
PSO and MLP	Optimized classificatio n of dimensional ity Data	*Concurr. Comput. Pract. Exp*	Wiley	2010	MLP,PSO	To prevent flooding and other risks arising from Rainfall	Recall, Precision, Accuracy
SVM in weather forecasting	Non linear regression models by utilizing the kernel functions	*Int. J. Comput. Theory Eng*	IACSIT Press	2009	MLP	To access the max. temperat ure	Accuracy

2.4 Conclusion

More than a hundred research journal papers from Science Direct, Scopus, and other resources are used to complete the literature survey to understand the various rainfall prediction models. There are some fit falls are observed in this way, and they are promoted to solve by developing proposed ensemble models. As shown in Table 2.4, the most relevant papers were used in the proposed research work. In the next chapter, discussed the system model, problem statement, and objectives of this research are designed.

Chapter 3

System Analysis and System Design

CHAPTER 3

SYSTEM ANALYSIS AND SYSTEM DESIGN

3.1 System Design Introduction

Chapter 2 presents the literature survey of various statistical and machine learning tools and techniques that are applied in analyzing and predicting rainfall. The chapter, covered, the role of different classification and regression tools to analyze weather reports and time series data in forecasting rainfall. This chapter presents the design hypothesis and mechanism of the proposed rainfall prediction system.

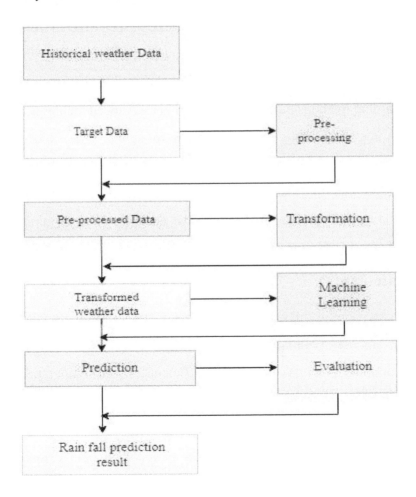

Fig. 3.1: General System model for Rainfall Prediction.

3.2 Problem analysis

Agriculture is the backbone of the Indian economy. Most Indian farmers primarily depend on seasonal rains for agricultural activities. Hence, rainfall prediction became so vital in the agricultural field. Rainfall and weather predictions enable the farmers in assessing crop production, decide the crop to be cultivated, predict the water resources required, save the crop from excess rainfalls, and so on. In recent times, many techniques were introduced to predict rainfalls in prior. Among these techniques, Machine Learning algorithms proved to be more helpful in forecasting rainfall [57].

A lot of research is done on varied machine learning algorithms such as Support Vector Regression, Genetic Programming, k-Nearest Neighbors, M5 Model trees, M5 Rules, and Radial Basis Neural Networks. The models are analyzed and tested extensively by utilizing the rainfall time-series dataset extracted from 42 cities, located in different regions of the Country such that, all different climatic features were covered in the dataset. The demonstrated results showed that the machine learning methods outperformed the other advanced techniques available. Another important motto of the work is to identify the hidden relations amid predictive accuracy and varied climates. In this regard, from the results, it is identified that the machine learning-based intelligent systems succeeded in predicting rainfall based on predictive accuracy with nominal correlations present among different climates [59].

3.3 Problem Statement

Weather conditions and rainfalls play a significant role in human life, especially in farmer's life. The amount of rainfall needed generally depends on plant type. For example, little or small amounts of water are sufficient for some plant types whereas, certain tropical plants require more water to survive. Hence, weather and rainfall prediction plays a very vital role in the agricultural supply chain process starting from crop selection till it reaches the final consumers. A small disturbance occurring at one stage in the supply chain process may break the entire cycle which leaves the farmer with huge losses. So, Timely prediction of rainfall enables the farmers in doing agriculture and in maintaining horticulture stocks required in India's coastal area.

Too little or too much rainfall may harm crop production. Continuous low rainfalls may kill the crop leading to erosion whereas excessively wet weather may lead to dangerous fungus growth. Similarly, heavy rainfalls may lead to floods. So, weather and rainfall predictions done prior will ease the farmers in managing their regular agricultural activities. In recent times, forecasting the state of the weather became a point of attraction for many researchers. Generally, the type of crop or plants that grow in a particular region primarily depends on the corresponding climatic conditions. It is, therefore, important to know the weather for the coming days to take precautions. Forecasting the future weather, especially rainfall won the attention of many researchers, as it enables the prevention of floods and other risks that arise from rainfall.

Accurate forecasting of Indian summer monsoon rainfall (ISMR) is highly desirable. Nevertheless, the scientific community is facing a lot of challenges in this area and is coming up with novel tools and techniques in solving them. Among the various scientific approaches available, numerical and statistical modeling gained so much popularity in recent times in weather and rainfall forecasting. However, soft computing failed in attaining such consideration, even though it is equal potential in handling the problem. Hence, it is extremely recommendable to assess the performance of the available soft computing techniques and to make the necessary improvements based on the acquired results.

The former studies were carried out by assessing the performance of varied machine learning and statistical methods in making rainfall predictions daily over a long period. In the proposed model, firstly, the occurrence and the intensity of the rainfall over a long period are considered. Further, the models are applied to the metrics and statistics collected to attain yearly, monthly as well as daily predictions. Finally, the results of all the models are analyzed and compared. The proposed model succeeded in attaining good accuracy with

39

minimal compute resources in predicting the rainfall. The collaborative approaches are used to prove that the proposed model is succeeded in predicting rainfalls with good accuracy. Multivariate and univariate analyses of variance are used to evaluate the differences among models.

3.4 Objectives

The main intent of this research work is to design and implement efficient rainfall prediction mechanisms by using machine learning Ensemble techniques.

The specific objectives of the research work are.

1. Apply Principal Components Analysis (**PCA**), Independent Component Analysis (**ICA**), and Linear Discriminant Analysis (**LDA**) techniques for Data pre-processing and feature enhancement.

2. Propose statistical models for rainfall prediction - Logistic regression, Decision tree, and Random Forest algorithms.

3. Propose Ensemble Machine learning predictive models such as **Cat boost** and **XG Boost** algorithms, for accurate rainfall prediction.

4. Develop a Weather prediction model using a hybrid approach based on **MLP** and **VAE** With firefly **optimization algorithm**.

3.5 Methodology

In India, over sixty percent of the population is either employed in agriculture or the agriculture industry. The agriculture sector has a significant role in the Indian economy. Even today, most Indian farmers depend on monsoon rains for their agricultural activities. Too much low or too much high rainfall may damage the crop leaving the farmers with huge losses and debts. Hence, effective rainfall prediction is very much important to overcome the problems as well as to enhance crop productivity.

The Proposed rainfall prediction system used the following process models to have time to time effective predictions:

1. Supervised learning dimensionality reduction techniques are used in data pre-processing and feature enhancement.

2. Statistical Machine learning Predictive Models such as Logistic Regression, Decision Trees, and Random Forest algorithms are used to generate timely rainfall predictions.

3. Ensemble Machine learning Predictive Models such as Cat boost and XG Boost algorithms are used to generate timely and accurate rainfall predictions.

4. A hybrid mechanism, the Fire-fly optimization algorithm is combined with VAE and MLP in implementing a novel Weather Prediction Model.

Supervised learning dimensionality reduction techniques are used in data Pre-Processing and Feature Enhancement. The data is initially pre-processed before it is submitted to prediction models. Data pre-processing involves the cleaning of data. In data pre-processing, initially, the null values as well as missing values are identified

And secondly, the identified values are handled thereby reducing the noise in the dataset. Further, the predictive model applies an embedded algorithm utilizing a Hybrid Mechanism based on Fire-fly Optimization Algorithm combined with VAE and MLP on the dataset to generate the time-to-time rainfall predictions.

3.6 System Architecture

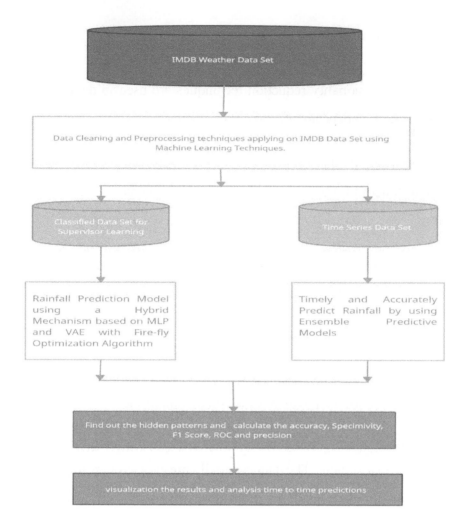

Figure 3.2: System Architecture for Rainfall Prediction.

The rainfall predictive system and its corresponding architecture are shown in figure 3.1 This system consists of the following six layers:

1. Database Layer: In this layer, the atmosphere data sets are extracted from the IMDB online, which are required to make rainfall predictions.

2. Data Cleaning and Pre-processing: In this layer, the data collected from the IMDB is pre-processed, analyzed, and further classified into data sets.

3. Classified data sets Layer: After pre-processing, the next step is to identify and classify the data as per hourly, monthly, and time-series

42

analysis.

4. Predictive Modeling Layer: The predictive machine Learning Models are applied to the data set to produce rainfall predictions of a specific location.

5. Results and Perception: This step enables identifying the hidden patterns and comparing the performances of varied models.

6. Visualizations: The results are analyzed and represented graphically using methods like visual graphs and so on.

3.7 Conclusion

By doing a depth study of the literature review and the existing work, the problem statement, research gaps, and different objectives are identified. Based on these problems, the proposed methodologies were developed, and the system architecture was designed. In the next chapter, discussed data preprocessing and features extraction techniques are implemented.

Chapter 4

Data Pre-Processing and Feature Enhancement

CHAPTER 4

DATA PRE-PROCESSING AND FEATURE ENHANCEMENT

4.1 Introduction

The efficiency of the supervised machine learning algorithms significantly depends on the quality of the data considered. So, it is mandated to pre-process the data before it is analyzed by a model. The most puzzling issue in analytic ML is identifying and removing noise values [R1]. In most of the cases, it is identified that the models are mainly deviating from attaining accuracy as the dataset possesses too many null values. These extremely deviating features are known as outliers. In addition, handling missing data values is another big challenge in data pre-processing steps [51].

In general, logical or symbolic learning algorithms are efficient at interpreting categorical data. But, the real-time datasets possess both numerical as well as label data elements [51]. Therefore, it is very difficult to discrete continuous data and to categorize the attributes of symbolic elements. The performance of the model can be enhanced with proper identification of the behavior of the data [51][58].

In general, real-time data possess several characteristics and features. But, only some of these features are of high priority on which the output function mainly depends. Hence, the data is primarily analyzed to find the important features as well as interdependencies among them. It also enables in identifying and removing outdated and redundant data. This step not only helps in decreasing the dimensionality of the data but also, makes the learning model faster and more effective [51].

4.2 Rainfall Data Set

The main objective of the thesis is to predict the probability of the occurrence of rainfall and its intensity. In India, over sixty percent of the population is either employed in agriculture or agriculture-based industries and even today, the majority of the Indian farmers depend on seasonal rains for their agricultural activities. So, the thesis focuses on implementing an accurate rainfall prediction system. To make

44

accurate predictions, the model has to be trained on real-time rainfall datasets. So, this section covers the information about the datasets and their corresponding attributes extracted from varied sources, www.imd.gov.in

This section mainly presents the dataset and its corresponding attributes. To extract the dataset, the data source used is IMD (Indian Meteorology Data). The dataset possesses almost 150000 records collected over a period between 1960 and 2019. The dataset consists of information about the rainfalls that occurred in the Visakhapatnam region on monthly basis. The attributes that are included in the dataset are Location, Minimum and maximum Temperatures(Degrees), Rainfall (Milli Meters), Sunshine(W/m^2), Evaporation(mm), Wind(Km/H), GustDir(m/s), Relative Humidity(g/kg), Cloud Coverage(oktas) and Pressure(pascal)[60].

The model implemented cannot be directly applied to the dataset, as the Sample dataset shown in Table 4.1, consists of both numerical and categorical data. Hence, before applying the model to the dataset, the dataset needs to be generalized and freed from noise.

Table 4.1: Description of sample data set and attributes.

S.No	Location	Max Temp (Degrees)	Min Temp (Degrees)	Rain Fall (MM)	Sun Shine (W/m^2)	Evapo-Ration (mm)	Relative humidity (g/kg)	Wind (Km/H)	Cloud (Oktas)	Pressure (Pascal)	Rain Today	Rain tomorrow
1	VSKP	22.9	13.4	0.6	10.6	5.8	71	44	8	1007.7	No	No
2	VSKP	30.1	13.1	1.4	8.3	7	58	28	NA	1007	Yes	No
3	VSKP	30.4	13.4	0	11.9	8	48	30	NA	1011.8	No	Yes
4	VSKP	21.7	15.9	2.2	5.9	6	89	31	8	1010.5	Yes	Yes
5	VSKP	32.3	17.5	1	5.1	4.8	82	41	7	1010.8	No	No
6	VSKP	29.7	14.6	0.2	5	2.6	55	56	NA	1009.2	No	No
7	VSKP	25	14.3	0	9.5	3.2	49	50	1	1009.6	No	No
8	VSKP	25.1	7.4	0	6	5.8	44	44	NA	1010.6	No	No
9	VSKP	25.7	12.9	0	9.4	4	38	46	NA	1007.6	No	No
10	VSKP	28	9.2	0	10.4	3.2	45	24	NA	1017.6	No	No
11	VSKP	26.7	7.7	0	9.7	7.2	48	35	NA	1013.4	No	No
12	VSKP	31.9	9.7	0	11	8.8	42	80	NA	1008.9	No	Yes
....	..	..	..	..	..	..	..	..	..	..	..	..

4.2.1 Data Set attributes

The following are the attributes that are included in the dataset:

MinTemp: Mintemp represents the minimum temperature in degrees Celsius.
MaxTemp: Max temp represents the maximum temperature in degrees Celsius.
Rainfall: It represents the amount of rain occurred per day and is measured in mm.
Evaporation: Measured using evaporimeters and is expressed in mm per day
Sunshine: Number of hours the bright daylight was there during a day.
WindGustSpeed: The highest wind gust recorded in 24 hours and is measured in km per hour.
RelHumid: The relative humidity in the air is considered to calculate the RelHumid.
Cloud: An eighth-sized cloud-obscured portion of the firmament oktas. It shows in what way the cloud covers a large portion of the sky. A 0 implies that there is no discernible heaven, while an eight indicates that the sky is entirely clouded.

Rain Today: It is a categorical binary label having label values that are either yes (1) or no (0). If rainfall exceeds 1mm in 24 hours (9 AM to 9 AM), then is represented with 1, otherwise 0

4.2.2 Identify and handle the missing values.

In real-world data sets, some of the values are absent due to various reasons like failure to load the information, corrupted data or data incompletely collected instrumental errors or manual errors, So, one of the crucial challenges faced by any data analyst is to handle such missing values as the models are trained and tested using these datasets. Let us look at different ways of correcting the missing values [117].

This technique suits best for the features, containing statistical data. In this approach, the incomplete data or missing values are replaced with the mean or median, or mode of the features' data. Instead of removing the entire row or column to handle the missing values, it is identified that this technique is generating better results. To handle the linear data, the technique, commonly used is to assess the deviation of neighboring values with it. Another numerical way to handle the missing values is to substitute the three estimates with those above mean, median, or mode [117].

1. Deleting the Rows

This is the most widely used technique in processing null values. If any certain feature possesses a missing values in the column's sample as shown in table 4.2, the specific row is deleted or a column is deleted if it possesses 70-75 % of missing values with corresponding mean/median.mode as shown in table 4.3. Before deleting the row or column, it is mandated to confirm that the deletion does not have any adverse impact on the results or outputs' accuracy [117].

Table 4.2. The missing values in the column's sample.

Location	Min Temp	Max Temp	Rainfall	Evaporation	Sunshine	Wind Gust Speed	Relative Humidity	Pressure	Cloud
VSKP	13.4	22.9	0.6	5.8	10.6	44	71	1007.7	8
VSKP	7.4	25.1	0	5.8	6	44	44	1010.6	6
VSKP	12.9	25.7	0	4	9.4	46	38	1007.6	4
VSKP	9.2		0	3.2	10.4	24	45	1017.6	7
VSKP	17.5	32.3	1	4.8	5.1	41	82	1010.8	7
VSKP	14.6	29.7	0.2			56	55	1009.2	2
VSKP	14.3	25	0	3.2	9.5		49	1009.6	1

Table 4.3. Replacing the missing values with corresponding mean/median/mode sample.

Location	Min Temp	Max Temp	Rainfall	Evaporation	Sunshine	Wind Gust Speed	Relative Humidity	Pressure	Cloud
VSKP	13.4	22.9	0.6	5.8	10.6	44	71	1007.7	8
VSKP	7.4	25.1	0	5.8	6	44	44	1010.6	6
VSKP	12.9	25.7	0	4	9.4	46	38	1007.6	4
VSKP	9.2	22.9	0	3.2	10.4	24	45	1017.6	7
VSKP	17.5	32.3	1	4.8	5.1	41	82	1010.8	7
VSKP	14.6	29.7	0.2	3.8	7.2	56	55	1009.2	2
VSKP	14.3	25	0	3.2	9.5	36	49	1009.6	1
Mean		22.9		3.8	7.2	36			

2. Predicting the Missing Values

Rather than removing the entire record to remove the missing values, the quantization technique predicts the missing value using the available data and replaces it with the predicted values. Once, all the missing values are substituted with the estimated values, the full dataset is examined using the same tools and techniques. The predicted values thus, generated are compared with the absolute value which is achieved after removing the entire cells and the standard deviation is analyzed. This technique is more advantageous compared to deleting the data item either row-wise or column-wise and precludes the standard distribution or deviation from being changed drastically [42][62].

3. Usage of algorithms supporting missing values

Another simple technique is to use machine learning algorithms to handle the missing or null values. KNN is a such popular machine learning algorithm that uses the principle of distance measurement. In this technique, KNN will consider the nearest k values to assess the missing or null value. For example, if the bus fares of an individual need to be assessed, then the bus fare of a similar person who matches him in his class, gender, and age will be considered [118].

4.2.3 Categorical Data Extraction

In general, to handle the categorical data, as shown in fig table 4.4 a perfect encoding process is needed to build a set of dummy values or variables corresponding to every value in the feature considered by using varied encoding techniques [118].

Table 4.4: Sample data with class label values

Location	Min Temp	Max Temp	Rain fall	Sunshine	Evap-oration	Relative humidity	Wind Gust Speed	Cloud	Pressure	Rain Today	rain tomorrow
VSKP	22.9	13.4	0.6	10.6	5.8	71	44	8	1007.7	No	No
VSKP	30.1	13.1	1.4	8.3	7	58	28	NA	1007	Yes	No
VSKP	30.4	13.4	0	11.9	8	48	30	NA	1011.8	No	Yes
VSKP	21.7	15.9	2.2	5.9	6	89	31	8	1010.5	Yes	Yes
VSKP	32.3	17.5	1	5.1	4.8	82	41	7	1010.8	No	No
VSKP	29.7	14.6	0.2	5	2.6	55	56	NA	1009.2	No	No

Label encoding is the process of assigning an integer value to every categorical variable value. For example, if a dataset possesses a categorical variable that contains the values from the set {"NO, " YES"}, then the technique, label encoding allot the numerical values that are derived from the set {0,1} correspondingly. The system generally deals with determined data only. Once, all the categorical values are mentioned, the system randomly assigns the numerical values to each categorical data correspondingly. An ordinal encoder as well as Mark Encoding assigns categorical variables' varied values to a definite integer value [63].

4.2.4 Splitting Data set as Training and Testing

The dataset extracted, generally serves two purposes. The first objective is to train the model as well as the second is to assess the model's performance. For this, the dataset is initially divided into two parts, training, and test datasets. The technique, to divide the extracted dataset into training and test datasets is also referred to as train-test split.

The train-test split technique is the most simple and popular procedure. The training dataset is utilized to train the model. Whereas, the test set is utilized to assess the model's performance. Firstly, the model is trained using the training dataset. Further, the trained model is applied to the test dataset. The outputs thus generated, are compared with the actual outputs, and based on the comparison results, the performance of the model is assessed. In some situations, especially when the size of the dataset is limited, the predictions can be made directly without the use of any model [44].

By each set of predictors, the definition of "significantly strong" is specific. It is recommended that an adequate dataset is needed to generate the training as well as test data sets. The specified training as well as test datasets need to be accurate enough to describe the actual problem [44].

In general, a dataset possesses thousands or millions of records. So, it is not possible to manually assess or identify all the interdependencies among the features. So, while the model is being trained, all the rare cases or special situations or hidden dependencies among the features will be identified [44].

Alternatively, if the dataset size is too small, it is not possible to use the train-test technique. Because, with a small number of records, it is not possible to train the model to assess the outputs for different input combinations. Similarly, when a small dataset is divided into training and test datasets, further, the size of the training and test datasets will be inadequate not only to train the model but also to assess the performance of the model. Hence, it also affects the reliability of the model. The approximations or predictions made by such a model will be either too optimistic or too pessimistic. So, an alternative approach available to train the model with small datasets is the k-fold cross-validation method. A configuration parameter is used to determine the size of the training as well as test datasets. The configuration percentage generally, lies between 0 and 1. For example, if the training set is derived with 0.63(63%) of the overall dataset, then the test set is derived from the remaining 0.37 (37%) of the dataset [44].

4.3 Feature Extracting Models

The sections above discussed the varied ways to handle the noisy data such as removing the null and missing values and what kind of dataset is needed for a predictive model. Section 4.2.4 covered the different techniques available in dividing the dataset into training and test datasets. Now, the next step is to build the appropriate features that are used by a predictive model in making effective and accurate rainfall predictions. Data

Feature extraction plays a vital role in making accurate predictions. The following section presents the varied techniques of feature extraction on the rainfall dataset.

4.3.1 Principal Components Analysis (PCA)

PCA (Principal Component Analysis) is the most popular data processing and data management technique. PCA technique is mainly used when the data possess a wide-range of variables that constantly change and are correlated positively. PCAs are a well-known dimension-reduction technique too. The motto is to decrease the size of the dataset. The technique initially finds the restricted set of features that have a high impact on the original output and removes the remaining features thereby reducing the dimensionality of the dataset with minimum data loss [46][65].

PCA's main steps as shown in fig 4.1 are as follows.

1. Choose and format the data in W (A voxels x 80 subjects). Here, each image view as a 1D column, so we have 80 columns and A rows data matrix.

2. Mean data in the center in W by average elimination.

3. Calculate the proprietary vectors.

The columns of Q reflect the proprietary vector of size A voxel, also called proprietary images. Self-data is stored as subcomponent data, a popular format for the storage of neuroimaging data. Column E reflects the associated own vectors of size 80 (equal to the number of subjects). Each vector E's coefficients are referred to as subject loads and represent each subject's contribution to the spatial pattern in the own corresponding data [46]. These PCA's steps explain how the PCA technique can be applied to the rainfall dataset to determine the correlation among the variables as well as in reducing the dataset dimensionality[64].

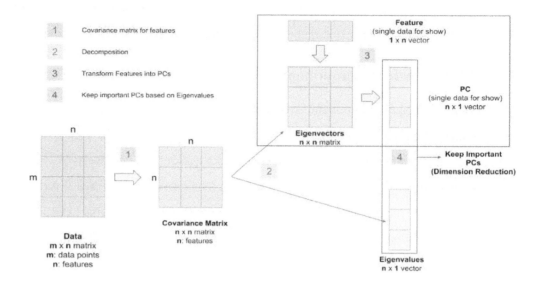

Fig 4.1: Principal Component Analysis steps and process.

$$s^2 = \frac{\sum_{i=1}^{n} (X_i - X)^2}{(n-1)}$$

Variance=Standard deviation^2

Table 4.5: Represents an attribute of temperature data.

X=Temperature	42	40	24	30	35	18	29	19	39

Table 4.6: Two dimensions of data temperature and humidity.

X=Temperature	40	28	37	18	25	17	29	38	30
Y=Humidity	90	70	90	80	60	70	90	80	70

Covariance: measures the correlation between X and Y

If Cov (X, Y) = 0   independent

Cov (X, Y) > 0   change in an identical way

Cov (X, Y) < 0   change in a contradictory way.

$$COV(X,Y) = \frac{\sum_{i=1}^{n}(X_i - \overline{X})(Y_i - \overline{Y})}{(n-1)}$$

More than two characteristics: Matrix of Covariance and Contains covariance values between all possible dimensions (attributes):

$$C^{nxn} = (c_{ij}|c_{ij} = cov(Dim_i, Dim_j))$$

Example for three attributes (x,y,z):

$$C = \begin{pmatrix} cov(x,x) & cov(x,y) & cov(x,z) \\ cov(y,x) & cov(y,y) & cov(y,z) \\ cov(z,x) & cov(z,y) & cov(z,z) \end{pmatrix}$$

Eigen values & eigenvectors

The eigenvectors of A are x vectors with the same direction as Ax (A is an n by n matrix). An Eigenvalue of A appears in the equation Ax=x.

$$\begin{pmatrix} 2 & 3 \\ 2 & 1 \end{pmatrix} x \begin{pmatrix} 3 \\ 2 \end{pmatrix} = \begin{pmatrix} 12 \\ 8 \end{pmatrix} = 4x \begin{pmatrix} 3 \\ 2 \end{pmatrix}$$

$$A\mathbf{x} = \lambda\mathbf{x} \Leftrightarrow (A-\lambda I)\mathbf{x} = 0$$

How to find x and: Calculate det(A-I), and you'll get a polynomial (degree n) Determine the roots to get det(A-I) = 0; the roots are the Eigenvalues. To get eigenvectors x, solve (A-I) x=0 for each. It indicates that we can obtain close to the maximum classification performance that can be attained with at least two or three original features with only one extracted feature. Rain will fall ('yes')/No Rainfall ('no') is two of the 11 qualities. There are 1, 50,000 occurrences, with 87,864 'yes' and 62,136 'no' and no missing data. We split the data into two sets, one for training and the other for testing. The proposed approach exhibits a considerable increase in performance over the previous method [64] [71].

Table 4.7: Data set representation.

Min Temp	Max Temp	Rainfall	Evapora-on	Sun shine	Wind Gust Speed	Relative humidity	Pressure	Cloud	Rain Today
13.4	22.9	0.6	5.8	10.6	44	71	1007.7	8	0
7.4	25.1	0	5.8	6	44	44	1010.6	6	0
12.9	25.7	0	4	9.4	46	38	1007.6	5	0
9.2	28	0	3.2	10.4	24	45	1017.6	6	0
17.5	32.3	1	4.8	5.1	41	82	1010.8	7	0
14.6	29.7	0.2	2.6	5	56	55	1009.2	7	0
14.3	25	0	3.2	9.5	50	49	1009.6	1	0
7.7	26.7	0	7.2	9.7	35	48	1013.4	5	0
9.7	31.9	0	8.8	11	80	42	1008.9	1	0
13.1	30.1	1.4	7	8.3	28	58	1007	9	1
13.4	30.4	0	8	11.9	30	48	1011.8	4	0
15.9	21.7	2.2	6	5.9	31	89	1010.5	1	1

The above table 4.7, represents the dataset generated after completing the pre-processing as well as encoding label classification. Where the categorical data with "yes" or "no" is encoded as 1 or 0.

The conversion of raw data into the inputs, required by a Machine Learning system is known as extracting the features. Its goal is to minimize the number of features in a dataset by recycling existing ones and creating new ones (and then discarding the original features). Its purpose is to reduce the number of features in a dataset by reusing existing ones to create new ones (and then discarding the original features). Metrics of Performance [66]:

- Accuracy = (TP+TN) / (TP+FP+FN+TN)
- Precision= TP/(TP+FP)
- Recall Or Sensitivity= TP/(TP+FN)
- F1 Score = 2*(Precision*Recall) / (Precision + Recall)

Classifier Accuracy Measures

Confusion matrix: A confusion matrix is a great tool for determining how effectively your classifier recognizes tuples from various classes.

Accuracy: The accuracy of a classifier is defined as the proportion of test set tuples successfully classified by it on a particular test set.

True positives: The term "true positives" refers to positive tuples that the classifier correctly labels.

True negatives: The classifier correctly identifies the negative tuples as true negatives.

False positives: Negative tuples that have been improperly labeled are known as false positives.

False negatives: Positive tuples that have been erroneously labeled are known as false negatives.

Macro average: It is used for evaluating performance score metrics (precision, recall, f1-score) for machine learning classification problems. It is simply the average of the precision of the different classes. If PC1, PC2, and PC3 are the precision of the respective classes. Then Macro average Precision= (PC1+PC2+PC3)/3.

Weighted average: It is a calculation that considers a variety of factors. [(rc1 * c1) + rc2 * c2)] = recall where rc1 and rc2 are the recalls for class1 and class2, and c1 and c2 are the number of instances in class1 and class2, respectively. When PCA

is applied to the aforementioned data set, the following confusion matrix and performance parameters are created as shown in the table 4.8 and table 4.9 respectively.

Table 4.8: PCA Confusion Matrix.

Classes	Rain Today=Yes	Rain Today=No	Total
Rain Today=Yes	81412(TP)	6452(FN)	87864
Rain Today=No	4048(FP)	58088(TN)	62136
Total	85460	64540	150000

Table 4.9: PCA Performance metrics.

	Precision (%)	Recall (%)	F1score (%)
0 Class	90	93	91
1 Class	95	92	93
macro avg.	92.5	92.5	92.5
Weighted avg.	92.9	92.4	92.6

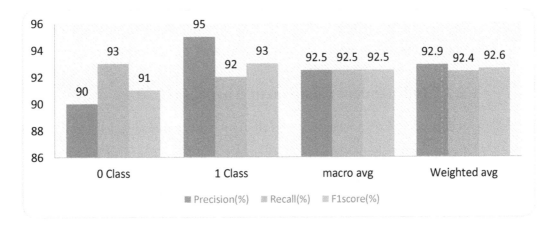

Fig 4.2: Precision, Recall, and F1 Score of PCA.

Fig 4.2 represents the 0 Class, 1 Class, and macro average and weighted average of performance metrics Precision, Recall, and F1 Score for PCA. The PCA performed well over the positive (1) class.

Table 4.10: Before and after extracted features of the PCA Performance

54

No of the features Selected out of 11 features	Before Extracted Features Classification Accuracy (%)	After Extracted Features Classification Accuracy (%)
1	63.2	91.4
2	74.2	92.5
3	84.5	93.5
10	94.5	95.0

The above table 4.10 represents before and after extracted features of the PCA Performance with the total of 11 selected features.

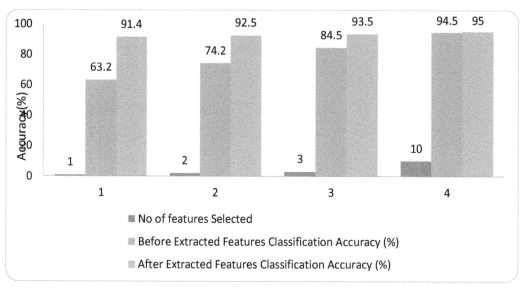

Fig.4.3: PCA Performance before and after features extraction.

From the above fig 4.3 it is observed that a drastic increase in performance by feature extraction. And also noticed it is proportional to the number of features increased. It is observed that PCA has given good results after feature extraction.

4.3.2 Independent Component Analysis (ICA)

Independent Component Analysis (ICA) is a technique in statistics used to detect hidden factors that exist in datasets of random variables, signals, or measurements. A typical independent component analysis example is the cocktail party problem. Herein, the cocktail party problem stands for its literal scenario that creates a noisy environment where people talking to each other cannot hear what the other person has said. The number of inputs registered by the algorithm is equal to the number of outputs produced. Herein, two assumptions are made before applying the technique of ICA. One, that the two signals (inputs) are 'statistically independent' of each other (unaffected by the occurrence of each other), and two, that these subcomponents are non-Gaussian (abnormal distribution) in nature.

What's more, the noisy environment also leads to a mixing of sound signals that, in turn, disables people from identifying the source of sound signals. This is where Independent Component Analysis steps in.

ICA is utilized to extract the important signals or data from the data source possessing a mixture of signals. The dataset may possess different types of data such as sounds or stock markets or videos. ICA is successfully being used in several applications such as audio signals, biological testing, and medical signals. ICA is also utilized in dimensionality reduction as it can preserve as well as delete a single source. ICA is even used as a filtering system as it is used both in filtering as well as erasing signals [R5]. ICA accepts a collection of individual components as input and deletes all the noise thereby defining each input correctly. Two features are considered to be independent if their linear, yet not linear influence is equal to zero [W3].

The basic steps in ICA are.

1. Preparing Data:

(a). Create a new dimensional input dataset (xT,c)T with N input features
x=[X1......XN]T and one output class c.

(b). Divide each feature fi by (fi-mi)/2i, where mi and 2i represent fi's mean and standard deviation, respectively.

2. Performance of ICA:

Apply ICA on the new dataset and save the (N+1) x (N+1) weight matrix W.

3. Shrinking small weights:

For each weight vector W_i of W, compute $a_i = \frac{1}{N+1}\sum_{j=1}^{N+1}|W_{ij}|$

For all W_{ij} in W, if $|W_{ij}| < \alpha.a_i$, then shrink $|W_{ij}|$ to zero, so here α is a small positive number.

Table 4.11: Confusion matrix of ICA.

Classes	Rain Today=Yes	Rain Today=No	Total

Rain Today=Yes	82535(TP)	5329(FN)	87864
Rain Today=No	3671(FP)	58465(TN)	62136
Total	86206	63794	150000

Table 4.11 and table 4.12,table 4.13 shows the confusion matrix of ICA and its performance metrics such as precision,recall and F1-score, After extracted features of the ICA Performance respectively.

Table 4.12: ICA Performance Metrics Precision, Recall, and F1-Score.

	Precision (%)	Recall (%)	F1score(%)
0 Class	91	94	92
1 Class	95	93	93
macro avg.	93	93.5	93.2
Weighted avg.	93	93	93

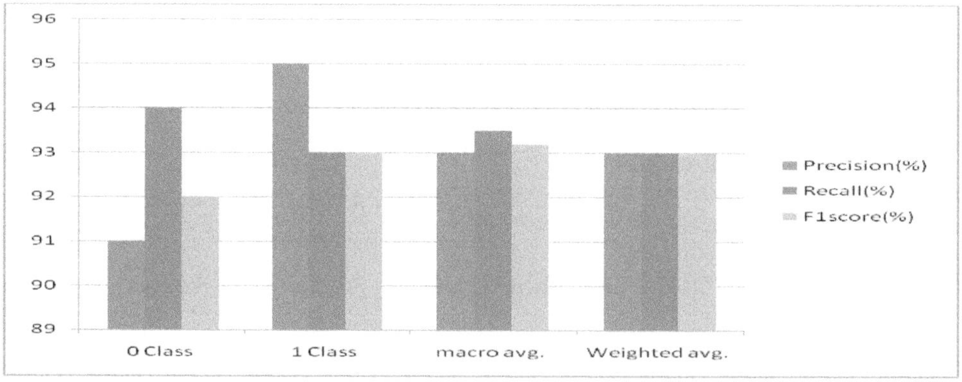

Fig.4.4: Precision, Recall, and F1 Score of ICA.

Fig 4.4 Represents the class 0/1 of performance metrics Precision, Recall, and F1 Score for ICA. ICA performed well over class 1 (positive) tuples.

Table 4.13: After extracted features of the ICA Performance.

No of features Selected	Before Extracted Features Classification Accuracy (%)	After Extracted Features Classification Accuracy (%)
1	89.2	93.4

2	91.2	94.5
3	93.5	95.5
10	94.5	95.8

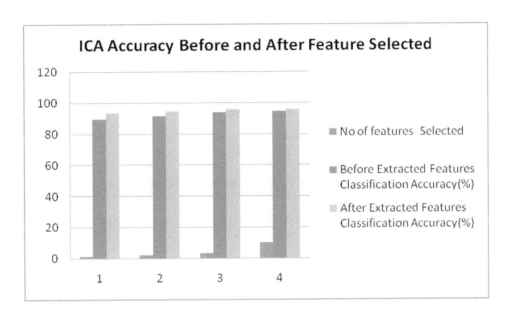

Fig 4.5: Before and after extracted features of the ICA.

It is observed from Fig 4.5, that ICA gives a small increase in performance before and after feature extraction. The number of features increased and ICA has consistent performance over the number of features selected.

4.3.3 Linear Discriminant Analysis (LDA)

LDA is a dimensionality reduction technique and uses supervised ML classifiers. This technique achieves classification by maximizing the distance amid each class's mean and minimizing the spread within the class. LDA performs better classification by maximizing each class's reach that is, by projecting the data in a lower-dimensional space. Therefore, LDA uses measures within as well as among categories. [118].

LDA Methodology and Steps

The following work presents the role of dimensionality reduction and feature engineering methods on the ML algorithms' performance in assessing the rainfall dataset [74].

58

Fig 4.6: LDA approach diagram.

The following are the general steps to perform a linear discriminant analysis (LDA) as shown in fig 4.6

1. Process the d-dimensional mean vectors for the different classes utilizing the dataset.

2. In sync two, compute the dissipate lattices (in the middle class and inside class disperse framework).

3. Compute the eigenvectors (e1,e2,..., ed) and relate eigenvalues for the dissipate frameworks (1,2,...,d).

4. To build a dk-dimensional lattice W, sort the eigenvectors by diminishing eigen values and select the k eigenvectors with the biggest eigen values (where each segment addresses an eigenvector).

5. Utilizing this dk eigenvector network, change the information into the new subspace. Y=XW (where X is a n , d-dimensional network encoding the n tests and y are the changed over n , k-dimensional examples in the new subspace) is an equation for changing over n, k-dimensional examples to another subspace. Grid duplication that sums up this [69][76].

Table 4.14: LDA Confusion Matrix.

Classes	Rain Today=Yes	Rain Today=No	Total
Rain Today=Yes	83490(TP)	4374(FN)	87864

Rain Today=No	1626(FP)	60510(TN)	62136
Total	85116	64884	150000

Table 4.14 and table 4.15,table 4.16 shows the confusion matrix of LDA and its performance metrics such as precision,recall and F1-score, Before and after extracted features selection Accuracy of LDA.

Table 4.15: LDA Performance Metrics Precision, Recall, and F1 Score.

	Precision(%)	**Recall(%)**	**F1score(%)**
0 Class	**93.2**	**97.2**	**95.1**
1 Class	**98**	**95**	**96.4**
Macro avg.	**95.6**	**96**	**95.7**
Weighted avg.	**96**	**95.9**	**95.8**

Table 4.16: Before and after extracted features selection Accuracy of LDA.

No of the features Selected	Before Extracted Features Classification Accuracy (%)	After Extracted Features Classification Accuracy(%)
1	73.4	92.3
2	85.6	93.5
3	92.5	95.5
10	95.5	96.2

60

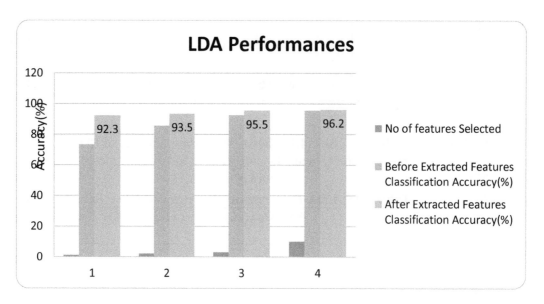

Fig 4.7: LDA Performance before and after features extraction.

It is observed that a drastic increase in performance by feature extraction. LDA gave a very good performance i.e. 96% overall. It worked well over both positive class and negative class tuples.

4.4 Results and Inferences

In this section, varied feature extraction methods are compared and analyzed. These models are developed using R programming and R Studio. The data set is extracted from the source, the IMD data provider. Predictive models cannot be applied directly to the extracted dataset as it possesses raw data. Hence, the feature extraction techniques such as PCA, ICA, and LDA and different dimensionality reduction techniques were applied to the dataset to make the dataset ready and to determine the accuracy, f1 score, Recall, specificity, and precision. The following results present the performance of the feature extraction models individually.

4.4.1 Accuracy of the feature extraction models.

In this section, the accuracy of the feature extraction models PCA, ICA, and LDA is calculated and assessed. Accuracy (ACC) is measured by dividing the number of accurate forecasts made by the total number of forecasts made. The range of accuracy lies between 0.0 to 1.0 [119].

$$ACC = \frac{TP + TN}{TP + TN + FN + FP} = \frac{TP + TN}{P + N}$$

True positive (TP): Correctly classified positive samples

False positive (FP): Incorrectly classified positive

samples.

True negative (TN): Correctly classified negative

samples

False negative (FN): Incorrectly classified negative

samples.

Table 4.17: Comparison of feature extraction algorithms accuracy.

S.no	Algorithm	Accuracy (%)
1	PCA	93
2	ICA	94
3	LDA	96

The above table 4.17 presents the accuracy attained by the feature extraction algorithms, PCA, ICA, and LDA correspondingly. The accuracy can be calculated by considering the number of true positive and true negative predictions the algorithm made divided by the total number of predictions. From the table, it is proved that LDA outperformed compared to PCA and ICA as it attained 96% accuracy whereas, PCA attained 93% and ICA attained 94% respectively.

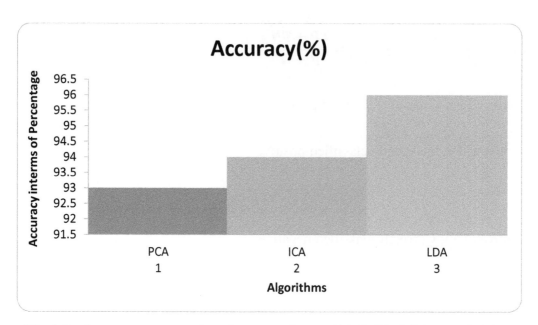

Fig 4.8: Accuracy comparison between the models PCA, ICA, and LDA.

Fig.4.8 represents the performance comparisons of the three feature extraction algorithms. The LDA performance is more compared to the other two models.

4.4.2 F1 score

It is nothing but a Correlation coefficient of the data set, the correlation between attributes with the classified data. The following represents the equation of the Mathew correlation [119][75].

$$MCC = \frac{TP.TN - FP.FN}{\sqrt{(TP + FP)(TP + FN)(TN + FP)(TN + FN)}}$$

4.4.3 Recall and Precision

The amount of accurate positive forecast is determined by Recall divided by a total of positive predictions. The highest sensitivity is 1.0, while the lowest sensitivity is 0.0.

Precision (PREC) tests the correct sum of successful predictions divided by

63

overall optimistic predictions. The maximum accuracy is 1.0, and the minimum accuracy of 0.0.

Table 4.18: Comparison of Precision.

Precision			
	PCA (%)	ICA (%)	LDA (%)
0 Class	91	95	96
1 Class	93	93	95
macro avg	92	94	96
Weighted avg	94	92	95

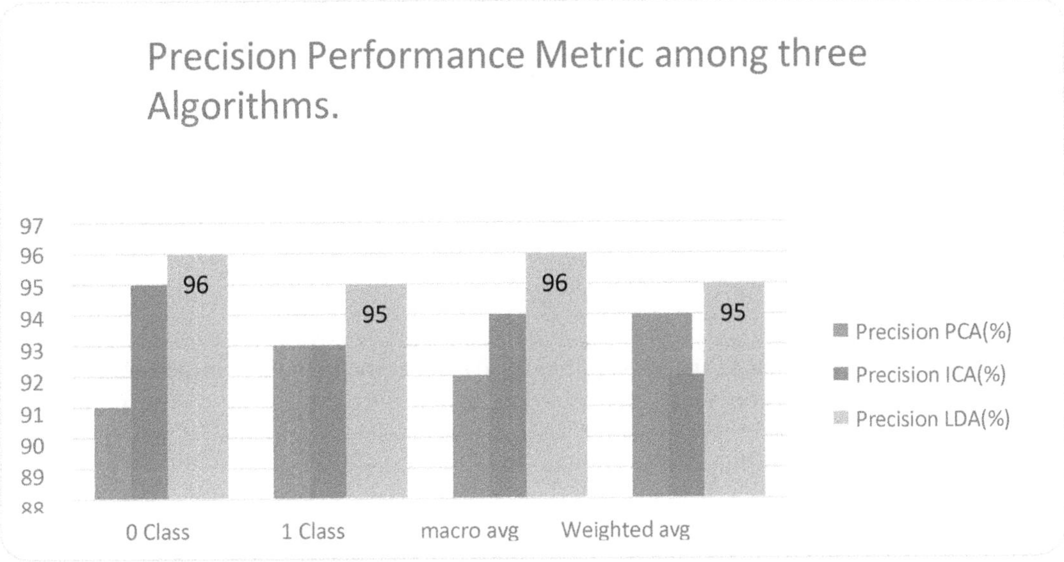

Fig 4.9: Precision performance of the class labels, macro, and weighed average.

Table 4.19: Attribute-wise performance.

	Min Temp	Max Temp.	Rainfall	Evaporation	Sunshine	Wind Gust Speed	Rel Humid	Cloud	Rain Today
F1 Score	92	90	86	93	86	95	88	93	93
Precision	87	96	92	88	95	88	89	87	97
Recall	97	97	89	90	87	90	96	96	92
Accuracy	88	90	92	88	86	88	88	88	86

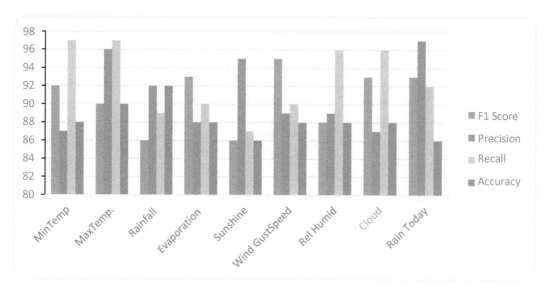

Fig 4.10: Attribute-wise performance.

4.5 Conclusion

The technique of Linear Discriminant Analysis (LDA) for data preprocessing performed well, over the other techniques in terms of precision, recall, and F1 score. The results showed fig 4.9 LDA algorithm is compared with Principal Components Analysis (PCA) and Independent Component Analysis (ICA). The next chapter discussed ensemble techniques for timely rainfall predictions.

CHAPTER 5

TIMELY AND ACCURATE PREDICTION OF RAINFALL BY USING ENSEMBLE PREDICTIVE MODELS

5.1. Introduction

The global climate is changing rapidly and constantly and accurate predictions are becoming a part of modern-day life. Varied economic sectors starting from farming to manufacturing and from tourism to transportation, mainly depends on atmospheric forecasts. As the biosphere is plagued by ongoing atmospheric changes and its output, it is extremely vital to accurately forecast the weather for smooth movement and comfortable daily activities. [46].

30 to 50 percent of the Indian wheat yields majorly depend on seasonal rainfall and the yield is affected by adverse weather conditions [53]. Incidentally, it is necessary to extract data from various meteorological centers for the application of ensemble models.

Statistical modeling is used to perform statistical analysis on the dataset and statistical models are used to visualize the data. It is observed that the data is analyzed as well as accurate predictions are derived when the data analysts utilized the mathematical models in their experiments. This approach enabled the data analysts in distinguishing between variables, make predictions of future data, and in visualizing the raw data [43].

Data analysts make use of appropriate models for predictions and analysis. The popular data analysis techniques are grouped into two categories: supervised learning, comprising classification and regression algorithms as well as unsupervised learning, comprising association and clustering techniques [42][56].

Ensemble Models are hybrid techniques that are formed by combining multiple models, either utilizing the same or varied algorithmic models, to build an application. Complex techniques like Bayesian, bagging, or boosting are used to enhance the results of the model.[31][77].

67

5.1.1 Ensemble method steps

1. Let K represents the training data, n denotes the total number of class labels, and J is the test data.
2. For a=1to n ;
3. Generate training set, Ki from K.
4. Derive base classifier Ci from Ki.
5. End
6. For each test record Y ∈ T ;
7. C* (Y) = vote(C1(K), C2(K),…….., Cn(K))
8. End.

5.2 Proposed Work

Climate and rain play a vital role in human life. Several sectors such as agriculture, harvesting, the food industry, and transportation depend on climatic conditions. All of these are procurement and sales processes. If the procurement process violates any point, eventually, the farmer will incur a loss. Accurate rainfall forecasts enable the farmers and other defenders in managing the activities in the need of coastal areas in India. To handle this, it suggests Ensemble Models (Catboost, XGBoost) as shown in fig 5.1. Most authors work to predict rain using mathematical models. It is extremely difficult to analyze as well as forecast a large amount of data using mathematical models. However, Ensemble models are possibly increasing the dynamics, and the use of prediction divisions makes it easier. This work presents the varied statistical models with ensemble models and showed in what way the characteristics of these algorithms are different from each other.

Here, the goal is to obtain weather predictions with accuracy. Nowadays, weather changes and rainfall data are vital. Depending on the rainfall, it is decided when to perform agricultural activities like plowing and cultivating. To generate accurate weather predictions, it is necessary to train the model using rainfall data comprising of a large amount of historical data. Further, the models are used in rainfall predictions. This function mainly addresses the multiplty of statistical

algorithms like a decision tree and random forest as well as Ensemble algorithms like

XGBOOST and CATBOOST. These algorithms are applied to datasets and the results proved, XGBOOST is the most effective algorithm when related to the rest of the algorithms discussed.

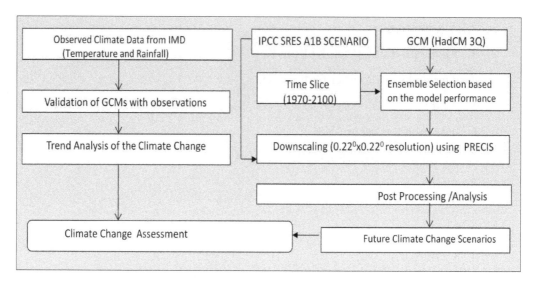

Fig 5.1: Climate change scenario with Ensemble models

For a particular dataset, a single algorithm may not be able to provide the optimal forecast. Machine learning algorithms have constraints, and creating a high-accuracy model is difficult. The overall accuracy of the model could be improved if we build and combine multiple models. The combination may be accomplished by combining the outputs from each model with two goals in mind: lowering model error while retaining generality. Some techniques can be used to implement this type of aggregation. These are also referred to as meta-algorithms [98].

5.3 System Architecture

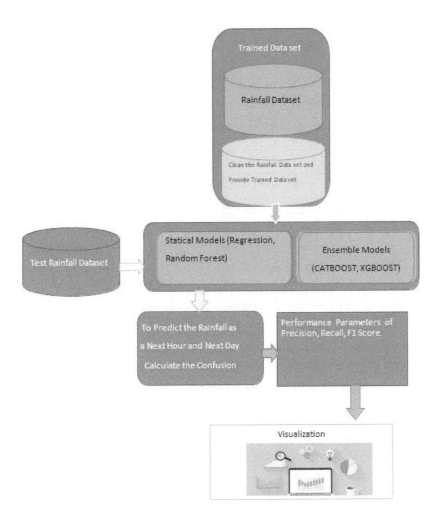

Fig 5.2: System Architecture

Fig 5.2 Describe the total prediction process of the rainfall. In this regard, total architecture consists the three major phases. The first one is the created rainfall data set. After setting up the rainfall training data set to apply to prediction modeling this one is the second phase of the architecture diagram. In the second phases train different regression algorithms like logistic regression algorithm, random forest, and ensembled models like CATBOOST, XGBOOST, etc. After thousands of records are trained to these prediction models the third phase begins. That is to test the real-time data set which contains the same attributes as the training data sets. After the testing process, the last phase is to calculate the algorithm performances in terms of accuracy, precision, recall, and F1 Score. These all algorithms are compared in terms of performance matrices to find out the best accuracy given algorithm these performance metrics are represented with visualization techniques.

70

5.4 Experimental Results

Table 5.1: Dataset sample and features.

Location	Min Temp	Max Temp	Rainfall	Sunshine	Evaporation	Wind Gust Speed	Pressure	Relative humidity	Cloud	Rain Today	Rain tomorrow
VSKP	22.9	13.4	0.6	10.6	5.8	44	1007.7	71	8	No	No
VSKP	25.1	7.4	0	6	5.8	44	1010.6	44	NA	No	No
VSKP	25.7	12.9	0	9.4	4	46	1007.6	38	NA	No	No
VSKP	28	9.2	0	10.4	3.2	24	1017.6	45	NA	No	No
VSKP	32.3	17.5	1	5.1	4.8	41	1010.8	82	7	No	No
VSKP	29.7	14.6	0.2	5	2.6	56	1009.2	55	NA	No	No
VSKP	25	14.3	0	9.5	3.2	50	1009.6	49	1	No	No
VSKP	26.7	7.7	0	9.7	7.2	35	1013.4	48	NA	No	No
VSKP	31.9	9.7	0	11	8.8	80	1008.9	42	NA	No	Yes
VSKP	30.1	13.1	1.4	8.3	7	28	1007	58	NA	Yes	No

The above table 5.1 shows the dataset sample and feautersand below table 5.2 represents the sample training data set with attributes in the expermintal results.

Table 5.2: Sample training data set with attributes.

Min Temp	Max Temp	Rainfall	Sunshine	Evaporation	Wind Gust Speed	Pressure	Relative humidity	Cloud
22.9	13.4	0.6	10.6	5.8	44	1007.7	71	8
25.1	7.4	0	6	5.8	44	1010.6	44	6
25.7	12.9	0	9.4	4	46	1007.6	38	5
28	9.2	0	10.4	3.2	24	1017.6	45	6
32.3	17.5	1	5.1	4.8	41	1010.8	82	7
29.7	14.6	0.2	5	2.6	56	1009.2	55	7
25	14.3	0	9.5	3.2	50	1009.6	49	1
26.7	7.7	0	9.7	7.2	35	1013.4	48	5
31.9	9.7	0	11	8.8	80	1008.9	42	1
30.1	13.1	1.4	8.3	7	28	1007	58	9
30.4	13.4	0	11.9	8	30	1011.8	48	4
21.7	15.9	2.2	5.9	6	31	1010.5	89	1

5.5 DATA EXPLORATION

Table 5.3: Data columns of non-null values.

S. No	Column	Non-Null
1	MinTemp	94773
2	MaxTemp	94773
3	WindGustSpeed	94773
4	WindDir9am	87119
5	Humidity9am	93482
6	Rainfall	93718
7	Evaporation	52579
8	Sunshine	94773
9	WindGustDir	94773
10	Cloud3pm	57476
11	Temp3pm	93514
12	Humidity3pm	93067
13	Pressure9am	85072
14	Cloud9am	94773
15	Rain Today	94773
16	Rain Tomorrow	87667

Table 5.3 presents the data exploration that possesses the column name and its related features. The above data reflects the sample 16 records out of a total of 94773 entries.

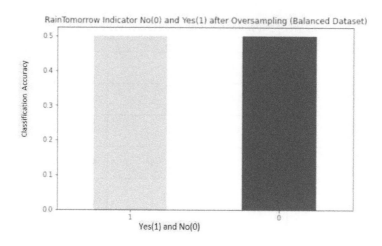

Fig 5.3: Class label value (Rain tomorrow = 0/1)

Fig 5.3 above bar chart describes the Classification of the data set in to Yes labeled and No Labeled in this graph shows the Classification of their percentages shown in the y-axis. And also Two bars represent the 1 classified count of the Light Blue Bar, and Also 0 represents dark blue

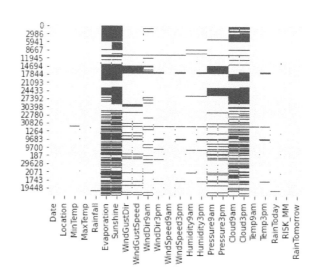

Figure 5.4: Trained dataset's Heat Map

Fig 5.4 Shows the data sets consisting of the values of the different attributes high, average, and min values. And Scatter plot represents with min value, max value, and average value display of all the attributes of the data set.

Table 5.4 presents the evaluation process of the Multiple Imputation by Chained Equations (MICE) packages. Further, the Inter-Quartile Range is used to find the outliers and to eliminate them forming the final working dataset.

Table 5.4: The final working dataset.

Date	1495.000000
Location	6.000000
Max-Temp	7.800000
Min-temp	8.400000
WindDir9am	8.000000
WindDir3pm	6.000000
Rainfall	2.600000
Sunshine	6.001871
Evaporation	3.600000
Wind Gust Dir	7.000000
WindSpeed9am	13.000000
WindSpeed3pm	11.536487
Wind Gust Speed	18.000000
Cloud9am	4.039764
Cloud3pm	3.943048
Temp9am	7.400000
Temp3pm	7.400000
Humidity9am	24.000000
Humidity3pm	32.000000
Pressure9am	6.900000
Pressure3pm	6.900000
rain today	1.000000
rain tomorrow	1.000000
RISK_MM	5.800000

74

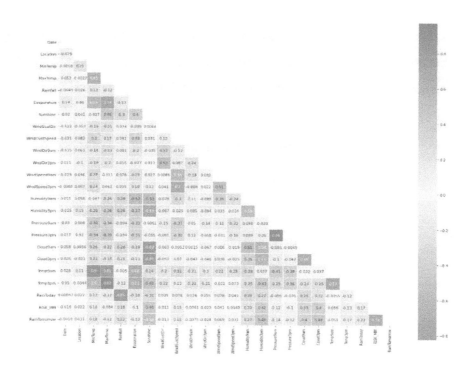

Figure 5.5: Correlations heat map in between Attributes

Fig 5.5 and 5.6 shows scatter plotting between the different attributes of the rainfall data sets it how the data is the behavior between the minimum to high values. Matric Graph of all the attributes of the data from the lowest level to the highest of the data.

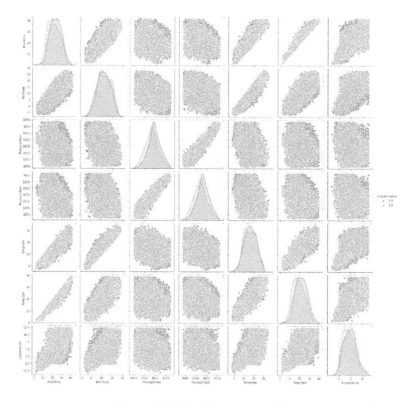

Figure 5.6: Attributes mapping.

75

5.6 Selection of Features

The section covers selecting the features as well as applying the enhancement techniques on the rainfall dataset. The features thus selected are:

{Date Location   MaxTemp   Mintemp   Evaporation
Rainfall,   Sunshine   Wind Gust Speed   Wind Gust Dir
WindDir9am   WindDir3pm   WindSpeed9am   WindSpeed3pm

Pressure9am   Pressure3pm   Humidity9am Humidity3pm Temp9am
Temp3pm   Cloud9am   Cloud3pm Rain Today   RISK_MM

Rain Tomorrow}

The corresponding sample values are as follows:

0	0.115284	0.03125	0.571429	0.518987	0.435873
	0.565373	0.468635	0.866667	0.521127	0.866667
	0.933333	0.516057	0.615385	0.539969	0.212121
	0.223214	0.277286	0.858482	0.409095	0.515957
	0.533627	0.174811	0.000000	0.0	
1	0.115575	0.03125	0.400000	0.574684	0.373192
	0.561494	0.727678	0.933333	0.521127	0.400000
	1.000000	0.108527	0.564103	0.236236	0.242424
	0.309524	0.297935	0.278259	0.223735	0.523936
	0.600252	0.174811	0.000000	0.0	
2	0.115866	0.03125	0.557143	0.589873	0.373192
	0.690262	0.775584	1.000000	0.549296	0.866667
	1.000000	0.490587	0.666667	0.168740	0.292929
	0.220238	0.324484	0.277728	0.167728	0.625000
	0.570937	0.174811	0.000000	0.0	
3	0.116157	0.03125	0.451429	0.648101	0.373192
	0.579307	0.756129	0.266667	0.239437	0.600000
	0.000000	0.286821	0.230769	0.247486	0.151515
	0.517857	0.445428	0.237588	0.173953	0.547872
	0.658881	0.174811	0.076923	0.0	
4	0.116448	0.03125	0.688571	0.756962	0.477660
	0.637367	0.384110	0.866667	0.478873	0.066667
	0.466667	0.184939	0.512821	0.663712	0.323232
	0.315476	0.244838	0.769158	0.670914	0.539894
	0.744161	0.174811	0.015385	0.0	

5.7 Logistic Regression Algorithm Implementation

Among the foremost well-known Machine Learning calculations, strategic relapse is used within the supervised Learning approach. It is a technique for associating degrees with an absolute dependent variable from a bunch of free factors. Calculated relapse is used to anticipate the results of a downright dependent variable. Thus, the result ought to seem as separate or all-out esteem. It alright is also affirmative or No, 0 or 1, valid or phony, etc, nonetheless, instead of careful qualities like zero and one, it conveys probabilistic qualities that square measure someplace within the middle. Calculated Regression is incredibly like simple regression, except for how it's used. Straight Regression is used to require care of relapse problems, though logistical Regression is used to handle grouping troubles. In calculated relapse, we tend to most extreme qualities utilizing an "S" shaped strategic capability as against a relapse line (0 or 1). The calculated capacity's bend portrays the chance of occasions, for instance, if cells square measure dangerous, if a mouse is stout enthusiastic about its weight, etc. Eq (5.1) Calculated relapse may be an important AI approach since it will provide chances and organize new data from each separate and constant dataset. Several styles of data may be accustomed to cluster perceptions utilizing calculated relapse[101].

$$lgo\left[\frac{y}{1-y}\right] = b_0 + b_1x_1 + b_2x_2 + b_3x_3 + \dots\dots\dots + b_nx_n \text{ Eq(5.1)}$$

Input to the algorithm:
{Max-Temp, Min-Temp, Evaporation, Rainfall, Sunshine, Wind Gust Speed, Wind Gust Dir, Pressure 9 am, Humidity9am, Temp3pm, Cloud9am, Cloud3pm, Rain Today, RISK_MM.}
Output: Prediction.

Table 5.5: Confusion matrix of Logistic Regression Algorithm.

Classes	Rain Today=Yes	Rain Today=No	Total
Rain Today=Yes	78098(TP)	13350(FN)	91448
Rain Today=No	21052(FP)	37500(TN)	58552
Total	99150	50850	150000

The above table 5.5 shows the Confusion matrix of Logistic Regression Algorithm and below table 5.6 represents performance of Logistic Regression Algorithm.

Table 5.6: performance of Logistic Regression Algorithm.

Parameters	Performances
Accuracy	77%
Cohen's Kappa	0.63
ROC area below Curve	0.51
Time Taken	25msec

Table 5.7: Performance metrics of Logistic Regression Algorithm.

	Precision (%)	f1-Score (%)	Recall (%)
Zero	73.1	74	85.4
One	78	70.4	64.3
macro avg	75.5	75.1	74.85
Average weight	77.2	77.1	77.6
Accuracy	74	77	75

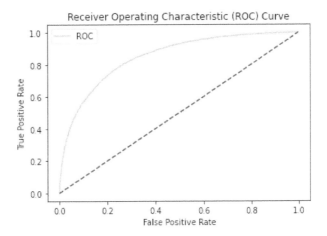

Figure 5.7: Operating characteristic Curve of Logistic Regression Algorithm.

From the above table 5.7 and fig 5.7, it is observed that the ROC area is moderate which indicates the Logistic regression is performed on the given dataset moderately i.e., it has an accuracy of 76%.

5.8 DECISION TREE

The decision tree is a tree-like structure used for decision-making. A decision tree is trained by segmenting the source set into subgroups. Recursive partitioning occurs when you apply the same method to each derived subset. When all of a node's samples belong to the same class, then the node is declared as a leaf node otherwise will continue the same procedure. A decision tree classifier is ideal for exploratory knowledge discovery. Big data is not a problem for decision trees. Eq. (5.2) The decision tree classifier has a high level of precision in many cases.

c

$$Entropy\ I_H = -\sum_{j=1} p_j \log_2(p_j)\ \ Eq.(5.2)$$

Basic Steps of Decision Tree Algorithm:

1. It can be designed in a top-down recursive divide-and-conquer procedure.
2. Initially, all training examples are considered as root.
3. All attributes are categorical.
4. Examples are divided recursively depending on the chosen attributes.
5. Test attributes are chosen, based on the statistical measure.

Conditions for stopping partitioning

1. The node samples belong to a single class.
2. No remaining attributes are left for further divisions – majority voting is used for classifying the leaf.
3. No samples left.

Input :
{MaxTemp, MinTemp, Evaporation, Rainfall, Sunshine, WindGustSpeed, WindGustDir, Pressure9am, Humidity9am, Temp3pm, Cloud9am, Cloud3pm rain Today, RISK_MM}.

Output: Prediction.

Table 5.8: Confusion matrix of Decision Tree Algorithm.

Classes	Rain Today=Yes	Rain Today=No	Total
Rain Today=Yes	78150(TP)	13669(FN)	91819
Rain Today=No	9900(FP)	48281(TN)	58181
Total	88050	61950	150000

Table 5.9: Performance of Decision Tree Algorithm.

Parameters	Performance
Accuracy	84%
ROC Area under Curve	0.84
Cohen's Kappa	0.72
Time Taken	34 msec

Table 5.10: Performance metrics of Decision Tree Algorithm.

	Precision (%)	Recall (%)	F1score (%)
0 Class	77.9	83	80.3
1 Class	88.7	85.1	86.8
Macro avg.	83.3	84	83.5
Weighted avg.	83.5	84.2	83.8

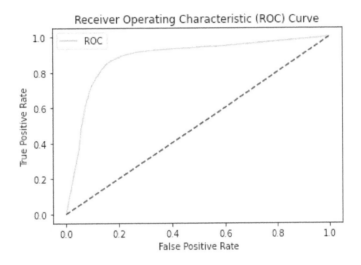

Fig 5.8: ROC Curve of Decision Tree

The above Tables 5.8,5.9 and 5.10 shows the performance metrics values of True False and False True, False True and False False values calculation and ROC Curve Calculations represents in Fig 5.8 of the Decision Tree Algorithm, These above diagrams represent the line and curve of the given specific data.

5.9 RANDOM FOREST ALGORITHM

A random forest is a machine-learning approach that may be used to address issues like regression and classification. It can be used for ensemble learning, which is an approach for solving complicated problems that involves multiple classifiers [28].

$$Random\ Forest = \frac{1}{N} \sum_{i=1}^{n} (fi - yi)^2$$

80

Here N is the total count of data points, fi is the model return value and yi is the original value of data point i.

5.9.1 The basic steps in Random Forest

Step-1: Choose randomly n data points from the training set.
Step-2: Construct decision trees, they are associated with the selected subsets.
Step-3: Select the N for decision trees that how many you required.
Step-4: Repeat Step 1 and 2.
Step-5: For the generated data points, find out the each decision tree prediction, and assign the generated points to the majority votes.

Input:

{MaxTemp, Mintemp, Evaporation, Rainfall,Sunshine,WindGustSpeed

WindGustDir ,Pressure9am , Humidity9am ,Temp3pm,Cloud9am ,Cloud3pm

RainToday,RISK_MM}

Output: Prediction.

Table 5.11: Random Forest Confusion matrix.

Classes	Rain Today=Yes	Rain Today=No	Total
Rain Today=Yes	78150(TP)	13669(FN)	91819
Rain Today=No	9900(FP)	48281(TN)	58181
Total	88050	61950	150000

Table 5.12: Random Forest Performance metrics.

	Precision (%)	Recall (%)	F1score (%)
0 Class	77.9	83	80.3
1 Class	88.7	85.1	86.8
Macro avg.	83.3	84	83.5
Weighted avg.	83.5	84.2	83.8

Table 5.13: Various parameters performance of Random Forest Algorithm.

Parameters	Performance
Accuracy	89%
ROC Area under Curve	0.84
Cohen's Kappa	0.79
Time Taken	28 msec

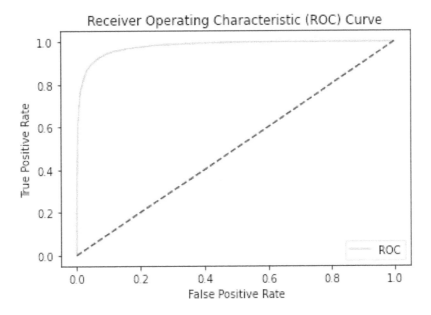

Fig .5.9: Random Forest ROC Curve.

The above Tables 5.11, 5.12, and 5.13 shows the performance metrics values of True False and False True, False True and False False values calculation and ROC Curve Calculations represents in Fig 5.9 of the Random Forest Algorithm, These all tables apply the data and find out the predictive the value and evaluate the performance metrics and draw the ROC curve display as tables and graphs.

5.10 Proposed CATBOOST Algorithm

Cat boost Decision tree (CBDT) is an emerging decision-boosting algorithm that can manage categorical features accurately. The following are some of the aspects in which CBDT stays different from conventional algorithms [101].

Input as a Categorical Data
Order data and find out the Number of Decision tress

82

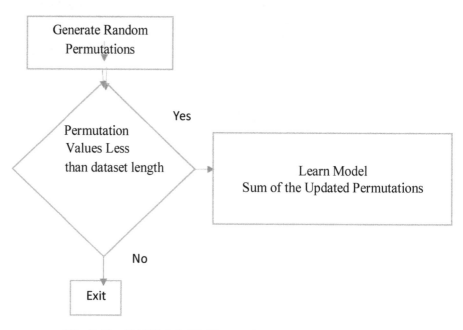

Fig 5.10: CATBOOST Flow Diagram.

The above fig 5.10 shows how the CATBOOST algorithm works as a flow chart. The following are the steps to working functionalities of the cat boost algorithm explained in detail. CATBOOST is working on the categorical data used to find the predictions. These predictions boost up on the categorical data [99].

5.10.1 Functionalities of CATBOOST

The training phase mainly deals with categorizing the features rather than involving in preprocessing activities. Cat boost requires the installation of all test datasets prior. Target statistics (TS) are the most efficient way to manage categorization with minimum loss of data. In reality, a random approval is generated by Cat boost for each event, including the standard tag number, for example, there is no change in the category value unless and until a new category is specified [102].

The characteristics of the whole category are new. It utilizes an inexplicable method to find compounds when building a new tree break. During the first break, it is not mandatory to think about the mix, whereas, during the splitting process for the second as well as subsequent times CATBOOST considers both the categorical functions as well as presets. The two categories opted as of the tree are considered as a two-value category and combined [109].

Distinguishing features develop with no refinement. TS process, distribution can vary and is used to convert the categorical features into numerical ones.

Quick scorer, CatBoost uses overlooked trees as basic predictions of the same separation parameters for the entire tree category. These are the modest trees that are rare in being overcrowded. Every leaf node of this forgotten tree represents a binary vector i.e. the length is calculated using the distance from the root node to the corresponding leaf node. This approach uses the CatBoost model as all binaries use encoded or statistical capabilities. [103].

Input : training set$\{(x_i, y_i)\}_{i=1}^n$ a different loss function L(y,F(x)),

Number 0f iterations M

Algorithm:

1. Initialize modal with a constant value:

$$F_0(x) = \arg \min \sum_{i=1} L(y_i, \gamma)$$

2. For m=1to M

1. Compute so-called pseudo-residual:

$$r_{im} = -\left[\frac{\partial L(y_i, F(x_i))}{\partial F(x_i)}\right]_{F(x)=Fm-1(x)} \qquad for\, i = 1 \ldots\ldots n.$$

Fit a Base Learner (e.g tree)$h_m(x)$ to pseudo-residuals, i.e train it using train set $\{(x_i, r_{im})\}^n$
$$\quad i=1$$

        3. Compute Multiplier γ_m by solving the following one-dimensional optimization problem:

$$\gamma_m = \arg \min \sum_{i=1}^{n} L(y_i, F_{m-1}(x_i) + \gamma h_m(x_i))$$

    4.Update the Modal

$$F_m(x) = F_{m-1}(x) + \gamma_m h_x(x)$$

4. Output $F_M(x)$

Input to the algorithm:

{   MaxTemp,MinTemp,Evaporation,Rainfall,Sunshine,WindGustSpeed WindGustDir ,Pressure9am , Humidity9am ,Temp3pm,Cloud9am ,Cloud3pm RainToday,RISK_MM}

Output: Prediction.

CATBOOST algorithms are used to make rainfall predictions. XGBOOST is an ensemble type algorithm used in rainfall predictions. The flow diagram and its functionality are as explained below [108].

According to CatBoost, it offers a good categorical data processing technique that supports numerical, categorical and text features. The CatBoost approach has a variety of choices for fine-tuning the processing step's properties. Gradient boosting machine learning is what CatBoost refers to as "boosting." Gradient boosting is a machine learning technique for dealing with regression and classification problems. This creates a prediction model from a set of weak prediction models, most often decision trees. Gradient boosting is an effective machine learning approach for finding solutions [67].

$$avg_{target} = \frac{countInClass + prior}{totalCount + 1}$$

Table 5.14: Confusion matrix of CATBOOST Algorithm.

Classes	Rain Today=Yes	Rain Today=No	Total
Rain Today=Yes	85500(TP)	4850(FN)	90350
Rain Today=No	9429(FP)	50221(TN)	59650
Total	94929	55071	150000

Table 5.15: Performance metrics of CATBOOST Algorithm.

	Precision (%)	Recall (%)	F1score (%)
0 Class	91.1	84.1	84.4
1 Class	90	94.6	92.2
Macro avg.	90.55	89.35	89.94
Weighted avg.	90.4	90	90.1

Table 5.16: Various parameters performance of CATBOOST Algorithm.

Parameters	Performance
Accuracy	90%
ROC Area under Curve	0.87
Cohen's Kappa	0.85
Time Taken	40 msec

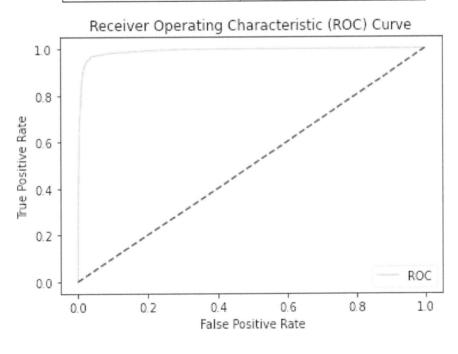

Fig 5.11: CATBOOST ROC Curve.

The above Tables 5.14, 5.15, and 5.16 shows the performance metrics values of True

False and False True, False True and False False values calculation and also ROC

Curve Calculations represents in Fig 5.11 of CATBOOST Algorithm.

5.11 Proposed XGBOOST Algorithm

"Extreme Gradient Boosting" is the acronym for XG Boost. XG Boost is a highly efficient, adaptable, and portable distributed gradient boosting library. XG Boost the Method is a gradient boosting algorithm, which is a prevalent approach in ensemble learning, as its name suggests.

XG Boost was first introduced in the year 2011 and continued to be polished and enhanced in later studies followed by Carlos Guestrin and Tianqi Chen. Paradigm is a learning method depending on the Boosting Tree model. The first derivative is alone used in traditional trees [20]. It is very difficult to train the nth tree as it depends on the initial n-1 trees. XG Boost uses CPUs multithreading process to runs achieves parallel loading and applies Taylor's second order extension. XG Boost uses different strategies to prevent over fitting [101].

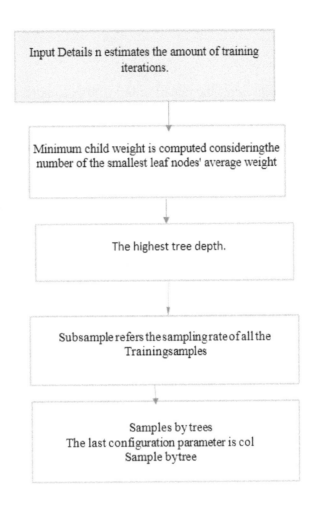

Fig 5.12: XGBOOST Algorithm Functionalities as a Flow Diagram.

5.11.1 XGBOOST Functionalities

In the proposed technique, the XG Boost regression model trains every target gene individually among the available 943 input genes. However, the XG Boost technique successfully avoids moderate overpasses and enhances the accuracy of the predictions [112][91].

The XG Boost design and performance are as below

1. Model completeness also depends on the number of estimators considered. Too low estimators will make the model incomplete. Similarly, too many estimators will make the model overflow.

2. Creates Min. Weight

The smallest leaf nodes are represented using the smallest boys' weight.

3. Limit Profiling

The ability and complexity of the tree model can be determined using the height of

88

the tree. Whereas simple algorithms are easy to overwrite.

4. under review

Under-review represents the rate of sample of the complete samples.

5. Examples

Sample tree coal is the final parameter considered. The sampling rate changes with the evolution of each tree.

6. Rate of learning

The learning rate boundary is a significant one that straightforwardly affects the calculation's presentation. By bringing down the heaviness of every individual movement, the model can suffer for an extensive stretch. As mentioned in the previous post, GBM separates the optimization problem into two phases by

first calculate the direction of the step and then optimizing the step length. XG Boost tries to determine the step directly by solving

$$\frac{\partial L(y, f^{(m-1)}(x) + f_m(x))}{\partial f_m(x)}$$

$$L(f_m)\alpha \sum_{j=1}^{Tm} [G_{jm}W_{jm} + \frac{1}{2}H_{jm}W_{jm}^2]$$

Then the loss function can be rewritten as:

$$L(f_m) \approx \sum_{i=1}^{n} \begin{array}{c} [g_m(x_i)f_m(x^i) + \frac{1}{2}h_m(x_i)f_m(x_i)^2] + const. \\ \alpha \sum_{j=1}^{T_m} \sum_{i \in R_{im}} [g_m(x^i)w_{jm} + \frac{1}{2}h_m(x_i)w_{jm}^2] \end{array}$$

A gradient boosting-based ensemble ML technique is implemented with XG Boost using decision trees. Artificial neural networks perform well in case of unstructured data and decision trees are suitable in case of unstructured data.

$$L^t = \sum_{i=1}^{n} l(y_i, y_i^{\wedge t-1} + f_t(x_i)) + \Omega(f_t)$$

y_i real value(label) known from the training dataset and $y^{\wedge t-1}$ cube seen as

Input:
{ MaxTemp,MinTemp,Evaporation,Rainfall,Sunshine,WindGustSpeed
WindGustDir ,Pressure9am , Humidity9am ,Temp3pm,Cloud9am ,Cloud3pm
RainToday,RISK_MM}

Output: Prediction.

Table 5.17: Confusion matrix of XGBOOST Algorithm.

Classes	Rain Today=Yes	Rain Today=No	Total
Rain Today=Yes	85500(TP)	4850(FN)	90350
Rain Today=No	2429(FP)	57221(TN)	59650
Total	87929	62071	150000

Table 5.18: Performance metrics of XG BOOST Algorithm.

	Precision (%)	Recall (%)	F1score (%)
0 Class	92.1	94.6	94.4

1 Class	97.2	95.9	94.2
Macro avg.	96.1	95.2	95.6
Weighted avg.	95	94.3	94.6

Table 5.19: performance of XG Boost Algorithm.

Parameters	Performances
Accuracy	**94%**
Cohen's Kappa	0.91
ROC area below Curve	0.89
Time Taken	32msec

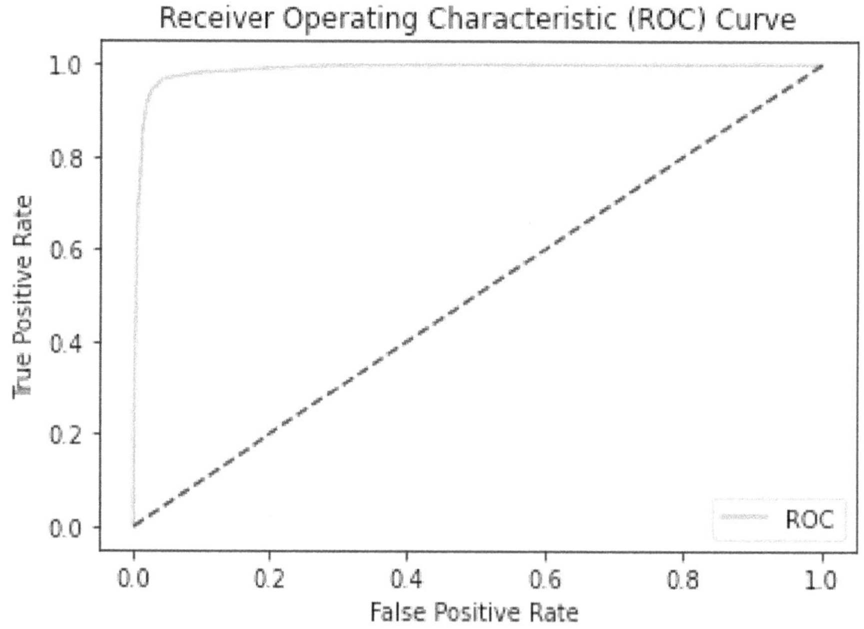

Fig 5.13: XGBOOST ROC Curve

The above Tables, 5.17, 5.18, and 5.19 Shows the performance metrics values of True False and False True, False True and False False values calculation and ROC Curve Calculations represents in Fig 5.13 of **XGBOOST** Algorithm

5.12 MODELS COMPARISON

This Section shows the Comparison of the algorithms respective of the time and Cohen's kappa values and finds out the best algorithm

91

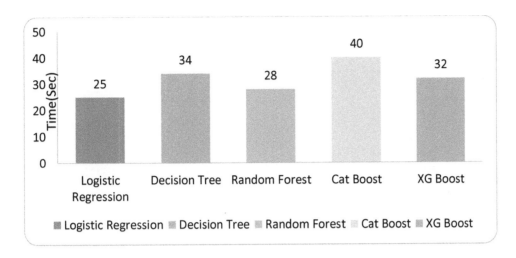

Fig .5.14: Model comparison over time taken.

In Fig 5.14 Bar Graph describes the performance metrics including time. In y Axis Time in milli seconds value to compare the logistic Regression, Decision Tree, Random Forest CAT Boost, and XG Boost Algorithms

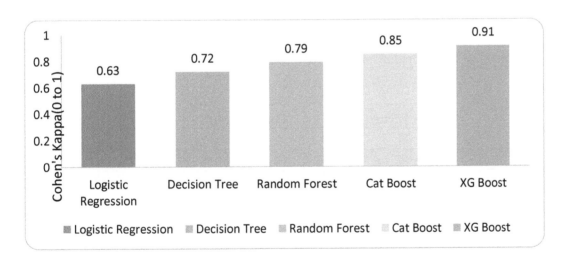

Fig.5.15: Models comparison over Cohen's Kappa constant.

Fig 5.15 Bar Graph describes the performance metrics including time. In y Axis Cohens Kappa 0 to 1 value to compare the logistic Regression, Decision Tree, Random Forest CAT Boost, and XG Boost Algorithms

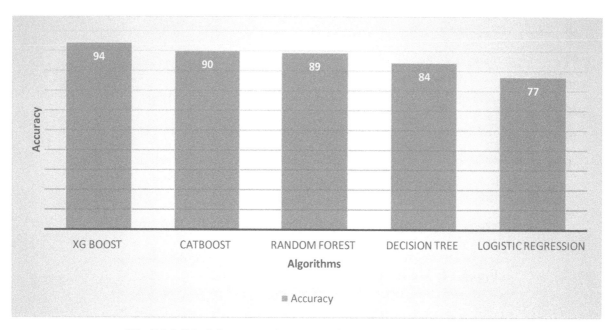

Fig.5.16: Models comparison over Accuracy constant.

Fig 5.16 Bar Graph describes the performance metrics including time. In y Axis Accuracy 0 to 100 percent value to compare the logistic Regression, Decision Tree, Random Forest CAT Boost, and XG Boost Algorithms

Table 5.20: Comparison of all algorithms with respective accuracy, time is taken, and Cohen's Kappa

Algorithm	Accuracy	Time Taken	Cohen's Kappa
XG Boost	94	32	0.91
CATBOOST	90	40	0.85
Random Forest	89	28	0.79
Decision Tree	84	34	72
Logistic Regression	77	25	0.63

5.13 Conclusion

The proposed ensemble algorithm XG Boost for weather prediction performed well, over other techniques in terms of accuracy, precision, recall, and F1 score. The results shown in fig 5.15 & fig 5.14, the XG Boost algorithm has better performance over LR, DT, RF, and CAT Boost algorithms. The next chapter discussed a novel weather prediction model based on MLP and VAE with a firefly optimization algorithm for accurate rainfall prediction.

93

Chapter 6

A Novel Weather Prediction Model using a Hybrid Mechanism based on MLP and VAE with Fire-fly Optimization Algorithm

Chapter 6

A Novel Weather Prediction Model using a Hybrid Mechanism based on MLP and VAE with Fire-fly Optimization Algorithm

6.1 Introduction

Government and private organizations belonging to various sectors uses weather predictions to improve the quality of their work and to diversify their daily activities. Weather predictions also enable the government to take necessary precautions during the natural disasters like floods and to protect people's health and property. Since the 1950s, the clarity of the forecast has increased rapidly with the advancements in technology, basic and practical sciences, and the introduction of emerging tools and methodologies. The accurate atmospheric features play a key role in generating efficient weather predictions [78].

Every year, billions of dollars are being spent by Governments and private companies in identifying accurate weather forecasts. Weather predictions are affecting many of the economic sectors both directly as well as indirectly. The probable sources of information related to climate are growing significantly. Recent advancements in machine learning enabling the government and private organizations to utilize this information in a better manner. Even though, it is not possible to derive hundred percent accurate weather predictions, but Artificial Intelligence tools are helping in enhancing the accuracy and resolution. Since, the weather predictions enable the government or organizations in predicting the heavy rain fall and floods in prior. Even the modest precautions that are taken can lead to big difference in economic loss. Even the emergency services like airports, farms, fire services and hospitals can be alerted with a day or at least some hours in prior[4].

Accurate weather predictions help in increasing the productivity of many industries such as agriculture, flight route of an airline and so on.

Perhaps the significant part of AI is to assess how it is incorporated through that of human behavior. Many researchers started observing how the weather conditions influences the choices we make like the food we take, places we visit, crop to be cultivated etc. Identifying the hidden patterns becomes easy with proper weather datasets. The accurate weather predictions allow the organizations in making proper and right decisions in prior [81].

At national and worldwide levels, much research has been done in agricultural as well as non-agricultural sectors on climate change depending on time-series data in a particular area. The time series can be categorized into linear and nonlinear datasets by using fusion models, which are formed by merging two or multiple computer models. Using hybrid model has been proven to be a better option compared over single models as it made predictions more accurately with respect to precision. On the contrary, the mixed methods also proved to be less accurate in some researches. So, ambiguities in the weather prediction models came up with new possibilities to develop a better and accurate forecast model. Many of the weather forecasters depend on time series data to predict the weather parameters. Inspired with sensory biological systems, a multi-line modeling tool, Feed-forward Neural Network is developed. It is a non-mathematical model. One of the key advantages is that the learning approach is made by finding similarities within different information and inputs without being able to connect between data entry and output [82]. In this research work, the Multi-Layer Feed forward Neural Network was selected as the ANNs architecture through an only unseen layer utilized to answer several difficult engineering problems [23]. The ANN is usually applied or done in two stages, namely, the preparation process and testing process. The first phase is training phase, where the sum of neurons in the hidden layer, the learning algorithm, the learning rate, activation mechanism and the momentum rate are computed by using Firefly algorithm. The second phase is testing phase, in which different learning algorithms are carried to find the optimal algorithm and to assess the functionality of the VAE-MLFNN approach.

A hybrid system as shown in fig 6.2, namely MLP (Multi-Layer Perceptron) combined with fire-retardant (VAE) process is been proposed. VAE (Variational Auto-Encoder) atmospheric data comprises of several features, among which some are national or worldwide whereas some are internal or domestic. It is not possible to extract or derive all such features utilizing a single method. Hence, a hybrid method is formed by merging VAE with MLP in order to extract and classify the weather features. VAE is used to extract the global characteristics and MLP is used to derive the internal features. Neural Networks are used to simulate the human nervous system. In addition, it is also proved that firefly optimization algorithm outperformed when compared over other widely used GA Genetic Algorithms) and PSO (Particle Swarm Optimization) techniques [26].

6.2 Proposed Work

Weather forecasts can have a profound effect on various sectors of society. Many governments as well as private organizations utilize these predictions to save human life, property and enhance the working processes and organize many people's daily activities. Since the 1950's, the accuracy of forecasting, that is, the dramatic increase in technology, basic and practical sciences, and the employment of emerging weather prediction approaches and methodologies has grown significantly. For making the predictions effectively at a specific place or region, accurate and advanced atmospheric parameters are required. Accurate rainfall forecasting has become one of the biggest concerns of water, and being aware of extreme weather conditions in prior will help in reducing the economic, human and property loss during the natural disasters.

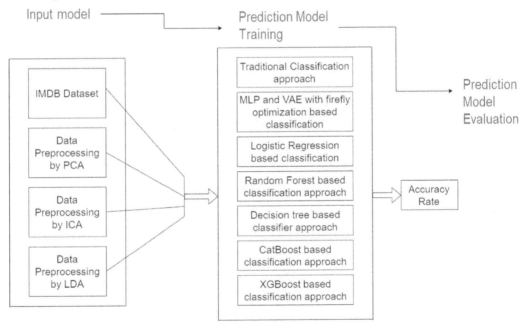

Fig. 6.1: Block diagram of Rainfall prediction models.

For researchers, the prediction method is built on the weather forecast in a variety of

97

fields as shown in fig 6.1, including machine learning [71][89],the weather data mining [68], mathematical predictions and functional hydrology [22]. Predictability is one of the biggest challenges. One of the most important concerns is how to interpret the current and historical observations in making accurate predictions. For short-term predictions, determination methods are sufficient to predict rainfall. Whereas, for long-term predictions, a hybrid model formed by combining VAE (variational auto-encoder) and MLP (Multi-layer perceptron) depending on the proposed firefly optimization process is required. As weather datasets possess several features, a single method may not be sufficient to analyze the features. Hence a hybrid process supported by MLP and VAE is utilized to obtain varied global as well as internal features. General or global features are extracted using VAE whereas; internal or local features are extracted using MLP.

The structure of the neural network consists of multi-layered neurons connected with each other. The vital part is to understand that a single layer of neural systems is helpful in developing AI [31][93].

Most of the Artificial Intelligence models are developing varied applications utilizing multi-layer display. ANNs (Artificial Neural Network) possessing multiple layers can be configured in a variety of ways. They usually contain a data layer which directs the weighted inputs through a sequence of hidden layers as well as a return layer towards the reverse direction. The systems require difficult setups and utilize sigmoid and varied techniques to regularize synthetic neurons. Partially, the systems are mechanically built, but many of them are made with the programming skills to design neural motion models [11][95].

6.3 Fire-fly Optimization Algorithm

The Fire-fly objective function is to be bringing varied fireflies as a flag frame. The measurement of the firefly can be computed as below according to [98][107]

1. The fireflies are usually drawn towards each other if both of them are unisexual.
2. Among both, switch to the most beautiful that is received.
3. If there is no beautiful firefly when compared to the specified firefly, then a random selection is done.

Firefly is measured to completely undermine the remaining transmission and

98

dedication to each of the most inspiring fireflies. Any change could happen during this phase, when the firefly decided to move closer to the magnificent firefly. Originally, it is worth expecting from the middle of the fireflies, w. Any type of travel rate considered in this case, before the Cartesian unit is ready for the entire case. The Cartesian distinction amid the two fireflies can be computed as:

$$Quv = t\,(Cur - Cvr) \text{ i.e } r=1$$

Here o_u represents the location vector of fire fly u and similarly, o_v represents the location vector of the fire fly corresponding with $o_u(l)$ that represents the first dimension's position value.

q Refers the consistency that is usually varies depending on whether two fireflies will depart or not. $q1$ is used to concentrate on the strength of the light. The four-sided imperative forms before $q_1(k) = i_0/k^2$ if the frequency of the firefly's activity intensifies. By this point the air accumulation, $q_2(k) = i_0/e^{(-wk)}$. In which, w belongs to the set $[0,\infty)$ and w is linked with the fireflies in a significant way.

In order to estimate and to avoid the ambiguity in implications, $q_1(0)$, we get $q(k) = i_0/e^{(-wk2)}$. Since, calculating exponential is costly. Instead of computing the value $i_0/e^{(-wk^2)}$, it is better to compute $i_0/(1+wk^2)$ as it costs less comparatively. Because, the look depends on the light power, $q_0 = i_0$, where q_0 belongs to the set $(0,1]$ and is the engaging quality at $k = 0$. The expression derived by combining the least two statements, we get $q(k) = q_0/(1 + wk^2)$. The change in the value of q_0 directly effects to what extent the fireflies are attracted towards each other. So, reducing the value of q_0 results in switching to lighter fireflies $o_i = o_i + q_0/(1 + wk_{xy}^2)(o_v - u) + pA(B - 0.5)$[105].

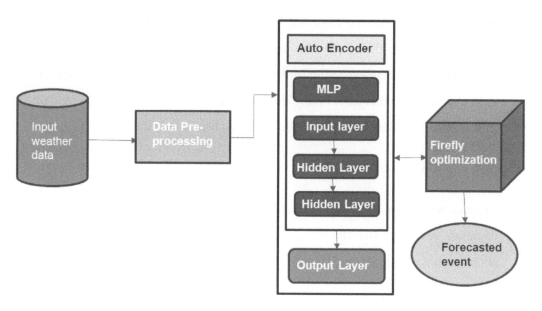

Fig. 6.2: The architecture of firefly algorithm combined with MLP and VAE.

6.3.1 Algorithm 6.1: Fire-fly Optimization

Input:
{MinTemp,MaxTemp,Rainfall
,Evaporation,Sunshine,Wind,GustSpeed,RelHumid,Cloud,Rain Today}
Output: Fireflies sorting
Result: Objective function $f(o)$ for $r-$ messing where $o = (o_1,o_2...o_r)$
The influence of light I_i on o_i is represented with $f(o_i)$;
Form p, s, q, w, and t.

Arbitrary places within y dimensions for N fireflies have;

Evaluation of complete N fireflies commence;

Though (final criteria not met)
; Resumed t ;

Fireflies for $u = 1$ to
num do
    Fireflies for $v = 1$
       to num do if Iv
       $> Iu$ then
       Relativize and switch firefly
       u to v; end

   end
Test current plans and restore light power;

   Make sure that the right location is identified or
not identified; end Lower the α ;
Fireflies sorting;

x represents the information number of somatic cell, The veiled layers' number of a somatic cell refers with y, and the yield number of the somatic cell is represented with z. In the layer the q the cell interface weight is represented as Kpq where p refers to set (1,2,...,x} and q refers to the set {1,2,...,y}. In addition, the weight corresponding to the nth somatic cell at unseen stage to the q considered from the set {1,2,...,y} and o considered from the set {1,2,...,k} is referred to as S no. NET invitation is in a secure layer of somatic cell as provided below [111].

$$A_q = X_p d_{pq} \quad \text{...............................} \quad (6.1)$$

$p = 1$ The trade work illustrates the exposed layers' motivation as:

$$E_q = G(A_q) = 1/((1+\exp r(-A\_q)) \quad \text{........................} (6.2)$$

The reward of NET c th neuron in the yield layer is referred

as: $NET_c = X_{eq}$ .... (6.3)

$q=1$

The equation 4 refers the activation function that reflects the process.

$$OUT_o = G(NET_o) = (1 + \exp(-NET_o) \quad \text{........................} (6.4)$$

6.3.2 FMLF2N2 Algorithm:

Input :{ Sorted Fly attributes of Weather attributes}

Output :{ prediction results with accuracy and F1 score, Precision, Recall} Result: Encoder and Decoder

Encoder Program;

A. RMSE which is calculated through equation 6.1 given by MLFNN;

B. Determine the light intensity, I_a for all the fireflies ;

   I_a, is in the context of wellness as RMSE was estimated in the simulation by the MLFNN of associated to $C_{a,1}, C_{a,2}, \& C_{a,3}$ figures for each firefly.; The ready data is a rummage-sale of MLFNN reconstruction during this phase. ; Created for m−people firefly,

C. **if** (RMSE(h) is larger than RMSE(l))

   Then Use equation 6.2 to adjust the h closing firefly l; **end**

D. Resulting to revitalize the firefly h conditions, by finding out $C_{h,b}$'s latest approximation for all k estimates ; Using equation 7, the RMSE is improved by the application of variable light concentration for h_m firefly. ; RMSE is also having reinvigorated characteristics for firefly h as new path into action of option components.;

E. Coordinate recurrence step-D before the citizens have identified absolute fireflies, and then h fireflies. ; The opposing firefly has been set to l on fire throughout each review. F. Set up the different n fireflies to be an h firefly. Accentuation organizes steps D and E until all of the fireflies are made of each other. ; Another social space is engulfed and registered as the general population for the next meeting throughout the whole of this point. ;

   F. The latest assembly in the climbing requirement of m fireflies offering the RMSE leg. The new better firefly will thus be designed for an overgrown reason.;

G. Coordinate recurrence of steps D, E, and F before the most exceptional level is a professional.;

An elegant firefly that renders the RMSE the least important is the finest overall firefly or the best game plan. ;

H. Excluding the estimate, the count of center points for covered component; Erudition and power levels have been obtained as parameters of the MLFNN from the best firefly overall.;

I. Conduct a clear research technique via the replication of test data of saved MLFNN; In addition, trade in the model's RMSE. ;

6.4 Experimental Results

Over a decade amid 2008 and 2017, Delhi climatic data was collected. The data thus extracted comprise of irregular or wrong values that can be usually removed during the preliminary data pre-processing phase. In a year, 365 weather records are generated and hence, for ten years 365 * 10 are generated. Each record in turn possesses several weather parameters such as wind speed, average temperature, moderate wind speed, humidity and moderate air pressure that occur in Delhi. Some of the events that occur can be no rain, rain, thunder, tornado, fog etc. Climate events are closely related to various atmospheric features like dust, temperature, humidity etc. Here, we mark the events of the decade amid 2008 and 2017. The proposed model is trained using the extracted dataset and further used to predict whether the climate event will be no rain, rain, storm, thunder, fog, hail and so on. The weather is now a days, intricate and the weather cannot be predicted easily in simple terms, as in the past [116].

Table 6.1: Support Vector Machine Algorithm Confusion Matrix.

Classes	Rain Today=Yes	Rain Today=No	Total
Rain Today=Yes	28769(TP)	1430(FN)	30199
Rain Today=No	9330(FP)	2840(TN)	12170
Total	38099	4270	42369

Table 6.2: Naive Bayes Algorithm Confusion Matrix.

Classes	Rain Today=Yes	Rain Today=No	Total
Rain Today=Yes	29456(TP)	1065(FN)	30521
Rain Today=No	4328(FP)	7520(TN)	11848
Total	33784	8585	42369

Table.6.3: MLP + VAE Algorithm Confusion Matrix

Classes	Rain Today=Yes	Rain Today=No	Total
Rain Today=Yes	31536(TP)	1130(FN)	32666
Rain Today=No	707(FP)	8996(TN)	9703
Total	32243	10126	42369

.

103

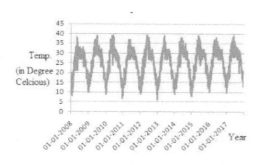 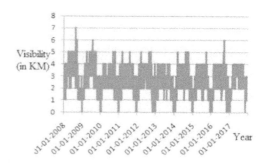

Fig. 6.3 (a) Temperature data extracted amid 2008-2017.

Fig.6.3 (b) Visibility data extracted amid 2008- 2017.

Distance visibility is computed by considering the path visible to a necked eye. Mist and dust fog blocks long-distance roads, sometimes in the morning until midday in winter. Climatic conditions will affect the appearance of the wind - low visibility causes adverse weather conditions, while great visibility leads to favorable weather [108].

Barometric pressure converts the local climate into a vital tool in order to make weather predictions. Areas with high barometric pressures are generally liable when compared to that are at low pressures. Even though, the exact barometric pressure value does not affect the atmospheric predictions, but the variation in barometric pressure affects the atmosphere significantly. The increase in barometric pressure in general leads to better weather conditions, while a decrease in wind leads to worse weather. Fig.6.4(a) represents the barometric pressure values in Delhi changing through b/w 994 and 1023, represented by. Fig 6.3(a) & 6.3(b) represents the data regarding the temperature and visibility collected over same period.

Humidity or moisture in air is another important feather extracted from weather dataset; Humidity is directly related to climatic conditions and rainfall. High humidity generally refers an increase in droplets in the air, resulting in soon rain. Low humidity content represents that the climate is dry. Values range from 13 to 100, extracted over a period ranging from 2008 to 2017.

Dew is another such important factor in assessing the climatic conditions. Especially, in the month of December, dew plays a vital role. Usually, the high impact of the weather and its performance in Delhi puts the first travel plan in Delhi. All different Routes to Delhi ranging from air to road, comprising flights, trains, as well as other official activities can be affected with climatic conditions. Presents dew in Delhi through the years ranging amid 2008 and 2017. In December, the Delhi's

104

average dew point is about -3. The temperature can reach its maximum, which are about 28 during the summer.

One of the most important factors that helps determine the weather is wind. It is the most difficult meteorological parameter in weather forecasting. It is especially important in winter as the temperature during winter is too low. Wind speed acts very vital especially when the temperature is below freezing. The Delhi wind speed is represented in, ranging from 0 to 40 for the period between 2008 and 2017.

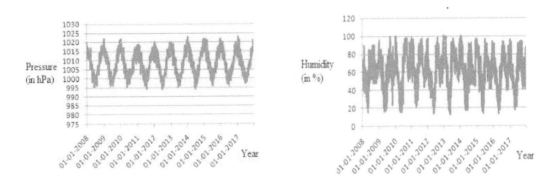

Fig.6.4 (a): Pressure values extracted amid 2008-2017.

Fig.6.4 (b): Humidity values extracted amid 2008-2017.

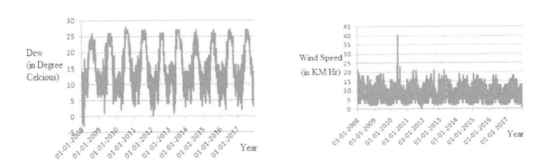

Fig. 6.5(a): Dew values extracted amid 2008-2017.

Fig. 6.5(b): Wind speed values extracted amid 2008-2017.

Delhi city daily rainfall over a decade is demonstrated below. It is identified that the rainfall through 2009 and 2014 is much better than in remaining years. Figure 6.6 represents the rainfall occurred with respect to individual months. There entire year is categorized into four groups: (1) January to February, (2) March to May, (3) June to September, and (4) October to December. From the graphs, it is undoubtedly showed that the rainfall is highest during the months July, August, and September. Generally, that time is the rainy season in India.

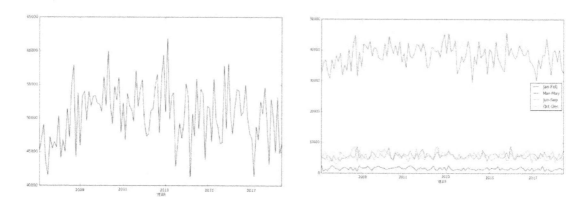

Fig 6.6: Monthly rainfall distribution over the period 2008-2017. The data is categorized into four (a) Daily rainfalls assessed over the period 2008-2017. Quarters :(1) January to February, (2) March to May,(3) June to September, and (4) October to December.

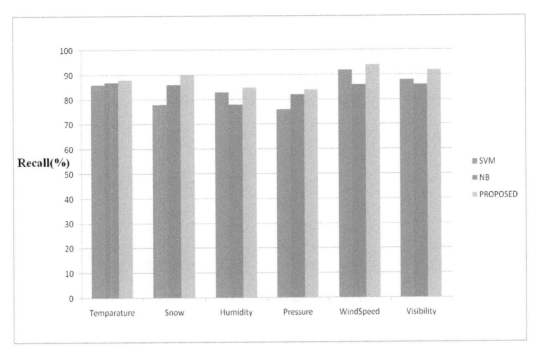

Fig 6.7: Recall performance comparison between SVM, NB and Proposed Algorithms.

The above fig 6.7 shows Performance analysis by Recall of weather data over last ten years to make classification of rainfall event based on atmospheric factors like temperature, wind speed, snow, humidity, pressure, etc...

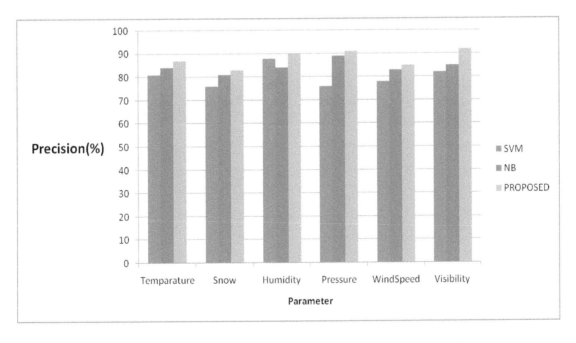

Fig. 6.8: Precision performance comparison between SVM, NB and Proposed Algorithms.

The above fig 6.8 shows the Performance analysis by Precision of weather data over last ten years to make classification of rainfall event based on atmospheric factors like temperature, wind speed, snow, humidity, pressure, etc...

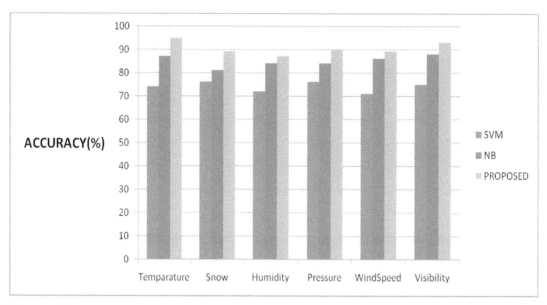

Figure 6.9: Accuracy performance comparison between SVM, NB and Proposed Algorithms.

The above fig 6.9 shows the Performance analysis by Accuracy of weather data over last ten years to make classification of rainfall event based on atmospheric factors like temperature, wind speed, snow, humidity, pressure, etc...

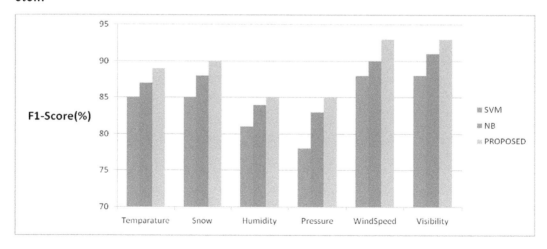

Fig 6.10: Comparing the models SVM, NB, Proposed model of F1- score.

The experimental results thus generated after a rigorous examination are discussed in detail. Presents a relative study of the precise values of the atmospheric data of the Delhi over the past decade. The proposed algorithm, MLP + VAE is used to categorize the rain events considering the climatic conditions such as wind speed, temperature, humidity, pressure, snow and so on. The values are analyzed using the proposed machine learning model MLP+VAE Combined with the firefly optimization process. Further, the experimental results are related with that of existing models of NB (naive Bayes) as well as SVM (Support Vector Machine). The results in fig 6.10 demonstrated that the values attained by utilizing the proposed technique are yielding better results when equated over the existing techniques in terms of accuracy.

In addition to the comparative value analysis of accuracy, an attempt is done to relate accuracy with recall values. It shows previous estimates of the Delhi climate over the past decade. The motto is to develop a rainfall prediction model based on atmospheric variables such as wind speed, temperature, rainfall, humidity etc. The proposed technique, MLP + VAE combined with a firefly optimization technique is analyzed and the experimental results attained are compared with the existing machine learning models like Naïve Bayes as well as Support Vector Machine.

Climate analysis in Delhi over the past decade has also been shown using estimates. The recall values thus attained are graphically represented in. The climatic parameters utilized in determining the rainfall classification like precipitation, temperatures, Rainfall, wind speed, humidity, and pressure etc. are demonstrated in Fig.6.9. Here, the extracted data is evaluated utilizing the firefly optimizer process that concentrates on machine learning based on MLP + VAE. The recall thus generated by the proposed model is compared with

109

SVM and NB and the comparative results proved that the proposed model is outperforming.

The climate analysis of the Indian capital over the past decade is represented utilizing the f1-score.As shown in F1 score is computed by considering the harmonic mean of precision and recall values. Represents the images of f1 attained from weather predictions. The rainfall event can be determined by considering the atmospheric variables like wind speed, temperature, precipitation, rainfall, humidity and pressure etc. Firefly optimization technique combined with MLP + VAE is used make the weather predictions. The f1-scores are computed using the proposed model and are further compared with Naïve Bayes as well as Support Vector Machine methods.   **6.5 Conclusion**
The proposed hybrid mechanism for weather prediction performed well, Over other techniques in terms of precision, recall and F1 score. The results are compared with SVM and Naive Bayes algorithms.

Chapter 7

Conclusion and Future Work

CHAPTER 7

CONCLUSION AND FUTURE WORK

7.1. Conclusion

The agriculture sector has a significant role in the Indian economy. In India, over sixty percent of the population is either employed in agriculture or the agriculture industry. Most Indian farmers primarily depend on seasonal rains for agricultural activities. Hence, rainfall prediction became so vital in the agricultural field. Climate change can be assessed by analyzing the changing conditions of weather over time. Especially, predicting the rainfall at a particular location needed certain features of weather such as wind speed, humidity, air direction, and so on. Many meteorological stations located at different places throughout the country are maintaining the corresponding regional weather data for many decades. The proposed model is analyzed on the weather dataset extracted from Indian Meteorological Department.

Climate change is a discipline involved with analyzing the varying distribution of weather for a specific period. Specifically, rainfall forecasting analyses specific features such as humidity and wind are used to predict rainfall for a specific location. This work mainly discusses statistical algorithms such as logistic Regression, random forest, and decision tree, and Ensemble algorithms CATBOOST and XGBOOST algorithms are applied for rainfall prediction. The motto of the work is to make predictions accurately. Weather datasets possess large amounts of data. This historical data is very crucial as the prediction model is trained using this data. Here, machine learning algorithms proved their excellence in training, analyzing, and making predictions on such large amounts of data. Research on Rainfall prediction happening for a long back, and accurate and timely prediction is a challenging problem. In this thesis, we attempted to improve the accuracy of prediction by using machine learning algorithms.

The Logistic Regression algorithm Classification accuracy is 77%, and Random Forest and Decision tree algorithms are 89% and 84% Classification accuracy respectively. In this regard, Ensemble algorithms Classification accuracy for CATBOOST is 90%

and XGBOOST is 94%. This research is carried out using a multi-layer Perceptron

111

combined with a variable auto-encoder (VAE). Further, the firefly optimization technique has been used to select suitable features. The experimental analysis proved that the proposed model is more effective than existing mechanisms. SVM and NB performances of the algorithm are 74.6%, and 87.5%, and the proposed algorithm Firefly Algorithm with MLP and VAE Accuracy is 95.6%.

7.2 Future Work

Many fields like agriculture, climate change, and the environment are prominent among them. Indian economy and growing country, directly and indirectly, depend on agriculture. In this present thesis, we have implemented ensemble machine-learning techniques for weather prediction. The work may be extended to prediction of heavy rainfall by applying the concepts of deep learning and artificial intelligence.

113

BIBLIOGRAPHY

[1]   M. S. Aakash Parmar, Kinjal Mistree, "Machine Learning Techniques for rainfall prediction: A Review," *Int. Conf. Innov. Inf. Embed. Commun. Syst.*, no. September 2017.

[2]   Abdul-Kader, H., & Mohamed, M. (2021). Hybrid machine learning model for rainfall forecasting. Journal of Intelligent Systems and Internet of Things, 1(1), 5-12.

[3]   L. Adhianto *et al.*, "HPCTOOLKIT: Tools for performance analysis of optimized parallel programs," *Concurr. Comput. Pract. Exp.*, vol. 22, no. 6, pp. 685–701, 2010, doi: 10.1002/cpe.

[4]   S. Aftab, M. Ahmad, N. Hameed, M. S. Bashir, I. Ali, and Z. Nawaz, "Rainfall prediction in Lahore City using data mining techniques," *Int. J. Adv. Comput. Sci. Appl.*, vol. 9, no. 4, pp. 254–260, 2018, doi: 10.14569/IJACSA.2018.090439.

[5]   R. Aguasca-Colomo, D. Castellanos-Nieves, and M. Méndez, "Comparative analysis of rainfall prediction models using machine learning in islands with complex orography: Tenerife Island," *Appl. Sci.*, vol. 9, no. 22, 2019, doi: 10.3390/APP9224931.

[6]   I. Ahmad, A. Singh, M. Fahad, and M. M. Waqas, "Remote sensing-based framework to predict and assess the interannual variability of maize yields in Pakistan using Landsat imagery," *Comput. Electron. Agric.*, vol. 178, no. July, p. 105732, 2020, doi: 10.1016/j.compag.2020.105732.

[7]   L. F. Ahyar, S. Suyanto, and A. Arifianto, "Firefly Algorithm-based Hyperparameters Setting of DRNN for Weather Prediction," *2020 Int. Conf. Data Sci. Its Appl. ICoDSA 2020*, 2020, doi: 10.1109/ICoDSA50139.2020.9212921.

[8]   R. Aler, R. Martín, J. M. Valls, and I. M. Galván, "A study of machine learning techniques for daily solar energy forecasting using numerical weather models," *Stud. Comput. Intell.*, vol. 570, pp. 269–278, 2015, doi: 10.1007/978-3-319-10422-5_29.

[9]   T. R. V. Anandharajan, G. A. Hariharan, K. K. Vignajeth, R. Jijendiran, and Kushmita, "Weather Monitoring Using Artificial Intelligence," *Proc. - Int. Conf. Comput. Intell. Networks*, vol. 2016-Janua, pp. 106–111, 2016, doi: 10.1109/CINE.2016.26.

[10]   V. Arcia, G. Corzo, and H. Calderón, "Active Moving Area Identification using Machine Learning. Case study: Ometepe Island , Nicaragua," p. 8462, 2021.

[11]   S. Aswin, P. Geetha, and R. Vinayakumar, "Deep Learning Models for the Prediction of Rainfall," *Proc. 2018 IEEE Int. Conf. Commun. Signal Process. ICCSP 2018*, pp. 657–661, 2018, doi: 10.1109/ICCSP.2018.8523829.

[12]   Rani, B. K., & Govardhan, A. (2013). Rainfall prediction using data mining techniques-a survey. Comput Sci Inf Technol, 3, 23-30.

[13]   W. Bao, N. Lianju, and K. Yue, "Integration of unsupervised and supervised machine learning algorithms for credit risk assessment," *Expert Syst. Appl.*, vol. 128, pp. 301–315, 2019, doi: 10.1016/j.eswa.2019.02.033.

[14]   Á. Baran, S. Lerch, M. El Ayari, and S. Baran, "Machine learning for total cloud cover prediction," *Neural Comput. Appl.*, vol. 33, no. 7, pp. 2605–2620, 2021, doi: 10.1007/s00521-020-05139-4.

[15]   Z. Chao, F. Pu, Y. Yin, B. Han, and X. Chen, "Research on real-time local rainfall prediction based on MEMS sensors," *J. Sensors*, vol. 2018, 2018, doi: 10.1155/2018/6184713.

[16]   D. Chauhan and J. Thakur, "Data Mining Techniques for Weather Prediction: A Review," *Int. J. Recent Innov. Trends Comput. Commun.*, vol. 2, no. 8, pp. 2184–2189, 2014, [Online]. Available: http://ijritcc.org/IJRITCC Vol_2 Issue_8/Data Mining Techniques for Weather Prediction A Review.pdf.

[17]   Y. J. Chen, E. Nicholson, and S. T. Cheng, "Using machine learning to understand the implications of meteorological conditions for fish kills," *Sci. Rep.*, vol. 10, no. 1, pp. 1–13, 2020, doi: 10.1038/s41598-020-73922-3.

[18]   D. Cho, C. Yoo, J. Im, and D. H. Cha, "Comparative Assessment of Various Machine Learning-Based Bias Correction Methods for Numerical Weather Prediction Model Forecasts of Extreme Air Temperatures in Urban Areas," *Earth Sp. Sci.*, vol. 7, no. 4, pp. 1–18, 2020, doi: 10.1029/2019EA000740.

[19]   S. Cramer, M. Kampouridis, and A. A. Freitas, "Decomposition genetic programming: An extensive evaluation on rainfall prediction in the context of weather derivatives," *Appl. Soft Comput. J.*, vol. 70, pp. 208–224, 2018, doi: 10.1016/j.asoc.2018.05.016.

[20]   S. Cramer, M. Kampouridis, A. A. Freitas, and A. K. Alexandridis, "An extensive evaluation of seven machine learning methods for rainfall prediction in weather derivatives," *Expert Syst. Appl.*, vol. 85, pp. 169–181, 2017, doi: 10.1016/j.eswa.2017.05.029.

[21]   A. Das, M. N. Khan, and M. M. Ahmed, "Detecting Lane change maneuvers using SHRP2 naturalistic driving data: A comparative study machine learning techniques," *Accid. Anal. Prev.*, vol. 142, no. October 2019, 2020, doi: 10.1016/j.aap.2020.105578.

[22] Y. Dash, S. K. Mishra, and B. K. Panigrahi, "Rainfall prediction for the Kerala state of India using artificial intelligence approaches," *Comput. Electr. Eng.*, vol. 70, no. May, pp. 66–73, 2018, doi: 10.1016/j.compeleceng.2018.06.004.

[23] Y. Dash, S. K. Mishra, S. Sahany, and B. K. Panigrahi, "Indian summer monsoon rainfall prediction: A comparison of iterative and non-iterative approaches," *Appl. Soft Comput. J.*, vol. 70, pp. 1122–1134, 2018, doi: 10.1016/j.asoc.2017.08.055.

[24] S. R. Devi, "Prediction of Rainfall Using Data Mining Techniques School of Electronics Engineering," *2018 Second Int. Conf. Inven. Commun. Comput. Technol.*, no. Icicct, pp. 1507–1512, 2018.

[25] J. Diez-Sierra and M. del Jesus, "Long-term rainfall prediction using atmospheric synoptic patterns in semi-arid climates with statistical and machine learning methods," *J. Hydrol.*, vol. 586, no. March 2020, doi: 10.1016/j.jhydrol.2020.124789.

[26] W. Dong, Q. Yang, and X. Fang, "Multi-step ahead wind power generation prediction based on hybrid machine learning techniques," *Energies*, vol. 11, no. 8, 2018, doi: 10.3390/en11081975.

[27] J. Elder, *The Apparent Paradox of Complexity in Ensemble Modeling* *, Second Edi. Elsevier Inc., 2018.

[28] A. Fouilloy *et al.*, "Solar irradiation prediction with machine learning: Forecasting models selection method depending on weather variability," *Energy*, vol. 165, pp. 620–629, 2018, doi: 10.1016/j.energy.2018.09.116.

[29] A. T. Fullhart, M. A. Nearing, R. P. McGehee, and M. A. Weltz, "Temporally downscaling a precipitation intensity factor for soil erosion modeling using the NOAA-ASOS weather station network," *Catena*, vol. 194, no. April 2020, doi: 10.1016/j.catena.2020.104709.

[30] I. Gad and D. Hosahalli, "A comparative study of prediction and classification models on NCDC weather data," *Int. J. Comput. Appl.*, vol. 0, no. 0, pp. 1–12, 2020, doi: 10.1080/1206212X.2020.1766769.

[31] A. Geetha and G. M. Nasira, "Data mining for meteorological applications: Decision trees for modeling rainfall prediction," *2014 IEEE Int. Conf. Comput. Intell. Comput. Res. IEEE ICCIC 2014*, pp. 0–3, 2015, doi: 10.1109/ICCIC.2014.7238481.

[32] M. U. Gutmann, "Un cor rec t ed Pro o Un cor rec t Pro o," 2017.

[33] J. T. Hancock and T. M. Khoshgoftaar, "Survey on categorical data for neural networks," *J. Big Data*, vol. 7, no. 1, 2020, doi: 10.1186/s40537-020-00305-w.

[34] S. E. Haupt, J. Cowie, S. Linden, T. McCandless, B. Kosovic, and S. Alessandrini, "Machine learning for applied weather prediction," *Proc. – IEEE 14th Int. Conf. eScience, e-Science 2018*, pp. 276–277, 2018, doi:

116

10.1109/eScience.2018.00047.

[35]   Hewage, P., Behera, A., Trovati, M., Pereira, E., Ghahremani, M., Palmieri, F., & Liu, Y. (2020). Temporal convolutional neural (TCN) network for an effective weather forecasting using time-series data from the local weather station. Soft Computing, 24, 16453-16482.

[36]   P. Hewage, M. Trovati, E. Pereira, and A. Behera, "Deep learning-based effective fine-grained weather forecasting model," *Pattern Anal. Appl.*, vol. 24, no. 1, pp. 343–366, 2021, doi: 10.1007/s10044-020-00898-1.

[37]   Huang, G., Wu, L., Ma, X., Zhang, W., Fan, J., Yu, X.& Zhou, H. (2019). Evaluation of CatBoost method for prediction of reference evapotranspiration in humid regions. Journal of Hydrology, 574, 1029-1041.

[38]   S. M. M. I. Diaz E. F. Combarro et al, "Machine Learning Applied to Weather Forecasting," *Springer*, vol. 15, no. 2, p. 99999, 2017.

[39]   L. Ingsrisawang, S. Ingsriswang, S. Somchit, P. Aungsuratana, and W. Khantiyanan, "Machine learning techniques for short-term rain forecasting system in the northeastern part of Thailand," *Proc. World Acad. Sci. Eng. Technol.*, vol. 31, no. 248–253, pp. 248–253, 2008.

[40]   S. Jain and D. Ramesh, "Machine Learning convergence for weather-based crop selection," *2020 IEEE Int. Students' Conf. Electr. Electron. Comput. Sci. SCEECS 2020*, 2020, doi: 10.1109/SCEECS48394.2020.75.

[41]   A. H. M. Jakaria, M. M. Hossain, and M. A. Rahman, "Smart Weather Forecasting Using Machine Learning:A Case Study in Tennessee," pp. 2–5, 2020, [Online]. Available: http://arxiv.org/abs/2008.10789.

[42]   W. Junjie, "CONSTRUCTION OF ATTENUATION RELATIONSHIP OF PEAK GROUND VELOCITY USING MACHINE LEARNING AND," pp. 3–6, 1995.

[43]   P. Kainthura and N. Sharma, "Machine learning techniques to predict slope failures in Uttarkashi, Uttarakhand (India)," *J. Sci. Ind. Res. (India).*, vol. 80, no. 1, pp. 66–74, 2021.

[44]   H. Kang, "The prevention and handling of the missing data," *Korean J. Anesthesiol.*, vol. 64, no. 5, pp. 402–406, 2013, doi: 10.4097/kjae.2013.64.5.402.

[45]   P. N. Kashyap, M. S. Preetham, R. V. Nayak, B. Uday, and T. V Radhika, "Smart Weather Prediction Techniques Using Machine Learning," *Int. Res. J. Mod. Eng. Technol. Sci.*, no. 07, pp. 1384–1389, 2020, [Online]. Available: https://irjmets.com/rootaccess/forms/uploads/smart-weather-prediction-techniques-using-machine-learning.pdf.

117

[46] F. Kherif and A. Latypova, "Principal component analysis," vol. 1, no. C, 1986.

[47] C. M. Ko, Y. Y. Jeong, Y. M. Lee, and B. S. Kim, "The development of a quantitative precipitation forecast correction technique based on machine learning for hydrological applications," *Atmosphere (Basel).*, vol. 11, no. 1, 2020, doi: 10.3390/ATMOS11010111.

[48] M. Kosanic and V. Milutinovic, "A Survey on Mathematical Aspects of Machine Learning in GeoPhysics: The Cases of Weather Forecast, Wind Energy, Wave Energy, Oil and Gas Exploration," *2021 10th Mediterr. Conf. Embed. Comput. MECO 2021,* 2021, doi: 10.1109/MECO52532.2021.9460245.

[49] B. Kosovic *et al.*, "A comprehensive wind power forecasting system integrating artificial intelligence and numerical weather prediction," *Energies*, vol. 16, no. 3, 2020, doi: 10.3390/en13061372.

[50] M. S. Koti and B. H. Alamma, *Big Data for Healthcare Databases.* Springer Singapore, 2019.

[51] Kotsiantis, S. B., Kanellopoulos, D., & Pintelas, P. E. (2006). "Data preprocessing for supervised leaning". International journal of computer science, 1(2), 111-117.

[52] V. M. Krasnopolsky and M. S. Fox-Rabinovitz, "Complex hybrid models combining deterministic and machine learning components for numerical climate modeling and weather prediction," *Neural Networks*, vol. 19, no. 2, pp. 122–134, 2006, doi: 10.1016/j.neunet.2006.01.002.

[53] M. Kuradusenge, S. Kumaran, and M. Zennaro, "Rainfall-induced landslide prediction using machine learning models: The case of ngororero district, rwanda," *Int. J. Environ. Res. Public Health*, vol. 17, no. 11, pp. 1–20, 2020, doi: 10.3390/ijerph17114147.

[54] M. A. Al Lababede, A. H. Blasi, and M. A. Alsuwaiket, "Mosques smart domes system using machine learning algorithms," *Int. J. Adv. Comput. Sci. Appl.*, vol. 11, no. 3, pp. 373–378, 2020, doi: 10.14569/ijacsa.2020.0110347.

[55] M. Lazri, K. Labadi, J. M. Brucker, and S. Ameur, "Improving satellite rainfall estimation from MSG data in Northern Algeria by using a multi-classifier model based on machine learning," *J. Hydrol.*, vol. 584, no. July 2019, 2020, doi: 10.1016/j.jhydrol.2020.124705.

[56] Lee, J., Lee, S., Hong, J., Lee, D., Bae, J. H., Yang, J. E., ... & Lim, K. J. (2021). Evaluation of rainfall erosivity factor estimation using machine and deep learning models. Water, 13(3), 382..

[57] W. Li, Y. Yin, X. Quan, and H. Zhang, "Gene Expression Value Prediction Based on XGBoost Algorithm," *Front. Genet.*, vol. 10, no. November, pp. 1–7, 2019, doi: 10.3389/fgene.2019.01077.

[58] Y. Li and Y. Qin, "The Response of Net Primary Production to Climate Change: A Case Study in the 400 mm Annual Precipitation Fluctuation Zone in China," *Int. J. Environ. Res. Public Health*, vol. 16, no. 9, 2019, doi: 10.3390/ijerph16091497.

[59] Z. Liu *et al.*, "Modelling of shallow landslides with machine learning algorithms," *Geosci. Front.*, vol. 12, no. 1, pp. 385–393, 2021, doi: 10.1016/j.gsf.2020.04.014.

[60] S. Manandhar, S. Dev, Y. H. Lee, Y. S. Meng, and S. Winkler, "A Data-Driven Approach for Accurate Rainfall Prediction," *IEEE Trans. Geosci. Remote Sens.*, vol. 57, no. 11, pp. 9323–9330, 2019, doi: 10.1109/TGRS.2019.2926110.

[61] N. Mishra, H. K. Soni, S. Sharma, and A. K. Upadhyay, "A comprehensive survey of data mining techniques on time series data for rainfall prediction," *J. ICT Res. Appl.*, vol. 11, no. 2, pp. 167–183, 2017, doi: 10.5614/itbj.ict.res.appl.2017.11.2.4.

[62] S. H. Moon, Y. H. Kim, Y. H. Lee, and B. R. Moon, "Application of machine learning to an early warning system for very short-term heavy rainfall," *J. Hydrol.*, vol. 568, no. November 2018, pp. 1042–1054, 2019, doi: 10.1016/j.jhydrol.2018.11.060.

[63] A. Mosavi, F. Sajedi-Hosseini, B. Choubin, F. Taromideh, G. Rahi, and A. A. Dineva, "Susceptibility mapping of soil water erosion using machine learning models," *Water (Switzerland)*, vol. 12, no. 7, pp. 1–17, 2020, doi: 10.3390/w12071995.

[64] G. Moshiashvili, N. Tabatadze, and V. Mshvildadze, "Jo u Pr pr oo," *Fitoterapia*, p. 104540, 2020, [Online]. Available: https://doi.org/10.1016/j.fitote.2020.104540.

[65] D. Muñoz-Esparza, R. D. Sharman, and W. Deierling, "Aviation turbulence forecasting at upper levels with machine learning techniques based on regression trees," *J. Appl. Meteorol. Climatol.*, vol. 59, no. 11, pp. 1883–1899, 2020, doi: 10.1175/JAMC-D-20-0116.1.

[66] K. Murugan, A. Seeta Reddy, M. Sivasainath Reddy, and P. Raviteja Reddy, "An Integrated Approach for Flood Prediction by Using Block Chain Network and Machine Learning," *IOP Conf. Ser. Mater. Sci. Eng.*, vol. 1049, no. 1, p. 012016, 2021, doi: 10.1088/1757-899x/1049/1/012016.

[67] S. Nalluri, S. Ramasubbareddy, and G. Kannayaram, "Weather prediction using clustering strategies in machine learning," *J. Comput. Theor. Nanosci.*, vol. 16, no. 5–6, pp. 1977–1981, 2019, doi: 10.1166/jctn.2019.7835.

119

[68] K. Namitha, A. Jayapriya, and G. S. Kumar, "Rainfall prediction using artificial neural network on map-reduce framework," *ACM Int. Conf. Proceeding Ser.*, vol. 10-13-Augu, pp. 492–495, 2015, doi: 10.1145/2791405.2791468.

[69] L. Naveen and H. S. Mohan, "Atmospheric weather prediction using various machine learning techniques: A survey," *Proc. 3rd Int. Conf. Comput. Methodol. Commun. ICCMC 2019*, no. Iccmc, pp. 422–428, 2019, doi: 10.1109/ICCMC.2019.8819643.

[70] S. Neelakandan and D. Paulraj, "An automated exploring and learning model for data prediction using balanced CA-SVM," *J. Ambient Intell. Humaniz. Comput.*, vol. 12, no. 5, pp. 4979–4990, 2021, doi: 10.1007/s12652-020-01937-9.

[71] V. B. Nikam and B. B. Meshram, "Modeling rainfall prediction using data mining method: A bayesian approach," *Proc. Int. Conf. Comput. Intell. Model. Simul.*, pp. 132–136, 2013, doi: 10.1109/CIMSim.2013.29.

[72] K. Palanivel and C. Surianarayanan, "an Approach for Prediction of Crop Yield Using Machine Learning and Big Data Techniques," *Int. J. Comput. Eng. Technol.*, vol. 10, no. 3, pp. 110–118, 2019, doi: 10.34218/ijcet.10.3.2019.013.

[73] Solomatine, D. P., & Shrestha, D. L. (2009). "A novel method to estimate model uncertainty using machine learning techniques". Water Resources Research, volume-45, issue-12.

[74] B. T. Pham *et al.*, "Development of advanced artificial intelligence models for daily rainfall prediction," *Atmos. Res.*, vol. 237, no. January 2020, doi: 10.1016/j.atmosres.2020.104845.

[75] S. Poornima and M. Pushpalatha, "Prediction of rainfall using intensified LSTM based recurrent Neural Network with Weighted Linear Units," *Atmosphere (Basel).*, vol. 10, no. 11, 2019, doi: 10.3390/atmos10110668.

[76] D. Prangchumpol and P. Jomsri, "Annual rainfall model by using machine learning techniques for agricultural adjustment," *J. Adv. Inf. Technol.*, vol. 11, no. 3, pp. 161–165, 2020, doi: 10.12720/jait.11.3.161-165.

[77] Praveen, B., Talukdar, S., Mahato, S., Mondal, J., Sharma, P., Islam, A. R. M., & Rahman, A. (2020). Analyzing trend and forecasting of rainfall changes in India using non-parametrical and machine learning approaches. Scientific reports, 10(1), 1-21.

[78] Prudden, R., Adams, S., Kangin, D., Robinson, N., Ravuri, S., Mohamed, S., & Arribas, A. (2020). A review of radar-based nowcasting of precipitation and applicable machine learning techniques. arXiv preprint arXiv:2005.04988.

[79] Y. Radhika and M. Shashi, "Atmospheric Temperature Prediction using Support Vector Machines," *Int. J. Comput. Theory Eng.*, vol. 1, no. 1, pp. 55–58, 2009, doi: 10.7763/ijcte.2009.v1.9.

[80] D. S. Rani, G. N. Jayalakshmi, and V. P. Baligar, "Low Cost IoT based Flood Monitoring System Using Machine Learning and Neural Networks: Flood Alerting and Rainfall Prediction," *2nd Int. Conf. Innov. Mech. Ind. Appl. ICIMIA 2020 - Conf. Proc.*, no. Icimia, pp. 261–267, 2020, doi: 10.1109/ICIMIA48430.2020.9074928.

[81] A. Raza *et al.*, "Comparative Assessment of Reference Evapotranspiration Estimation Using Conventional Method and Machine Learning Algorithms in Four Climatic Regions," *Pure Appl. Geophys.*, vol. 177, no. 9, pp. 4479–4508, 2020, doi: 10.1007/s00024-020-02473-5.

[82] G. T. Reddy *et al.*, "Analysis of Dimensionality Reduction Techniques on Big Data," *IEEE Access*, vol. 8, pp. 54776–54788, 2020, doi: 10.1109/ACCESS.2020.2980942.

[83] D. A. Sachindra, K. Ahmed, M. M. Rashid, S. Shahid, and B. J. C. Perera, "Statistical downscaling of precipitation using machine learning techniques," *Atmos. Res.*, vol. 212, pp. 240–258, 2018, doi: 10.1016/j.atmosres.2018.05.022.

[84] N. A. M. Salim *et al.*, "Prediction of dengue outbreak in Selangor Malaysia using machine learning techniques," *Sci. Rep.*, vol. 11, no. 1, pp. 1–9, 2021, doi: 10.1038/s41598-020-79193-2.

[85] A. G. Salman, B. Kanigoro, and Y. Heryadi, "Weather forecasting using deep learning techniques," *ICACSIS 2015 - 2015 Int. Conf. Adv. Comput. Sci. Inf. Syst. Proc.*, pp. 281–285, 2016, doi: 10.1109/ICACSIS.2015.7415154.

[86] N. Samsiahsani, I. Shlash, M. Hassan, A. Hadi, and M. Aliff, "Enhancing Malaysia Rainfall Prediction Using Classification Techniques," *J. Appl. Environ. Biol. Sci*, vol. 7, pp. 20–29, 2017.

[87] V. Sasikala, K. K. K. Venkata, S. Venkatramaphanikumar, P. A. Babu, M. E. Kumar, and N. G. Krishna, "Weather Predictive System using Machine Learning Algorithms," *J. Xi'an Univ. Archit. Technol.*, vol. XII, no. Vi, pp. 31–39, 2020, [Online]. Available: https://xajzkjdx.cn/gallery/44-june2020.pdf.

[88] S. Scher and G. Messori, "How Global Warming Changes the Difficulty of Synoptic Weather Forecasting," *Geophys. Res. Lett.*, vol. 46, no. 5, pp. 2931–2939, 2019, doi: 10.1029/2018GL081856.

[89] C. J. S. Sci, L. I. U. Siqing, C. Yanhong, H. U. Qinghua, and Y. Tianjiao, "Development of New Capabilities Using Machine Learning for Space Weather Prediction *," China National Knowledge infrastructure. vol. 40, no. 5, pp. 875–883, 2020, doi: 10.11728/cjss2020.05.875

121

[90]   J. H. Seo, Y. H. Lee, and Y. H. Kim, "Feature selection for very short-term heavy rainfall prediction using evolutionary computation," *Adv. Meteorol.*, vol. 2014, 2014, doi: 10.1155/2014/203545.

[91]   Sethi, N., & Garg, K. (2014). Exploiting data mining technique for rainfall prediction. International Journal of Computer Science and Information Technologies, 5(3), 3982-3984..

[92]   U. Shah, S. Garg, N. Sisodiya, N. Dube, and S. Sharma, "Rainfall prediction: Accuracy enhancement using machine learning and forecasting techniques," *PDGC 2018 - 2018 5th Int. Conf. Parallel, Distrib. Grid Comput.*, pp. 776–782, 2018, doi: 10.1109/PDGC.2018.8745763.

[93]   Shamshiband,S.,   Hashemi,   S.,Salimi,H,.Samadianfard,   S., Asadi,E.,Shadkani,S.,..&   Chau,K.W.(2020).   Predicting   standardized streamflow index for hydrological drought using machine learning models.Engineering   Applications   of   Computational   Fluid Mechanics,14(1),339-350.

[94]   O. Simeone, "A Very Brief Introduction to Machine Learning with Applications to Communication Systems," *IEEE Trans. Cogn. Commun. Netw.*, vol. 4, no. 4, pp. 648–664, 2018, doi: 10.1109/TCCN.2018.2881442.

[95]   N. Singh, S. Chaturvedi, and S. Akhter, "Weather Forecasting Using Machine Learning Algorithm," *2019 Int. Conf. Signal Process. Commun. ICSC 2019*, pp. 171–174, 2019, doi: 10.1109/ICSC45622.2019.8938211.

[96]   S. Singh, G. Noida, F. A. Nagrami, G. Noida, A. P. Pillai, and G. Noida, "Weather prediction by machine learning."

[97]   N. Solanki and G. Panchal, "A novel machine learning based approach for rainfall prediction," *Smart Innov. Syst. Technol.*, vol. 83, no. Ictis 2017, pp. 314–319, 2018, doi: 10.1007/978-3-319-63673-3_38.

[98]   N. Sundaravalli and A. Geetha, "A Study & Survey on Rainfall Prediction And Production of Crops Using Data Mining Techniques," *... Res. J. Eng. ...*, pp. 1269–1274,   2016,   [Online].   Available: https://www.academia.edu/download/54341943/IRJET-V3I12284.pdf.

[99]   A. Tharwat, "Independent component analysis: An introduction," *Appl. Comput. Informatics*, vol. 17, no. 2, pp. 222–249, 2018, doi: 10.1016/j.aci.2018.08.006.

[100]   C. Tortora, P. D. McNicholas, and F. Palumbo, "A Probabilistic Distance Clustering Algorithm Using Gaussian and Student-t Multivariate Density Distributions," *SN Comput. Sci.*, vol. 1, no. 2, pp. 1–22, 2020, doi: 10.1007/s42979-020-0067-z.

[101]   F. Tufaner and A. Özbeyaz, "Estimation and easy calculation of the Palmer Drought Severity Index from the meteorological data by using the advanced machine learning algorithms," *Environ. Monit. Assess.*, vol. 192, no. 9, 2020, doi: 10.1007/s10661-020-08539-0.

122

[102] R. UshaRani, T. K. R. Krishna Rao, and R. Kiran Kumar Reddy, "An Efficient Machine Learning Regression Model for Rainfall Prediction," *Int. J. Comput. Appl.*, vol. 115, no. 23, pp. 24–30, 2015, doi: 10.5120/20292-2681.

[103] G. Verma, P. Mittal, and S. Farheen, "Real Time Weather Prediction System Using IOT and Machine Learning," *2020 6th Int. Conf. Signal Process. Commun. ICSC 2020*, pp. 322–324, 2020, doi: 10.1109/ICSC48311.2020.9182766.

[104] V. A. Vuyyuru, G. Apparao, and S. Anuradha, "Timely and Accurately Predict Rainfall by using Ensemble Predictive Models," *IOP Conf. Ser. Mater. Sci. Eng.*, vol. 1074, no. 1, p. 012019, 2021, doi: 10.1088/1757-899x/1074/1/012019.

[105] B. Wang, P. Feng, C. Waters, J. Cleverly, D. L. Liu, and Q. Yu, "Quantifying the impacts of pre-occurred ENSO signals on wheat yield variation using machine learning in Australia," *Agric. For. Meteorol.*, vol. 291, no. November 2019, 2020, doi: 10.1016/j.agrformet.2020.108043.

[106] C. C. Wei and T. H. Chou, "Typhoon quantitative rainfall prediction from big data analytics by using the apache hadoop spark parallel computing framework," *Atmosphere (Basel).*, vol. 11, no. 8, 2020, doi: 10.3390/ATMOS11080870.

[107] J. A. Weyn, D. R. Durran, and R. Caruana, "Improving Data-Driven Global Weather Prediction Using Deep Convolutional Neural Networks on a Cubed Sphere," *J. Adv. Model. Earth Syst.*, vol. 12, no. 9, 2020, doi: 10.1029/2020MS002109.

[108] W. L. Woon, Z. Aung, and S. Madnick, "Data Analytics for Renewable Energy Integration: Second ECML PKDD Workshop, DARE 2014, Nancy, France, September 19, 2014, Revised Selected Papers," *Lect. Notes Comput. Sci. (including Subser. Lect. Notes Artif. Intell. Lect. Notes Bioinformatics)*, vol. 8817, pp. 81–96, 2014, doi: 10.1007/978-3-319-13290-7.

[109] J. Xia *et al.*, "Machine Learning-based Weather Support for the 2022 Winter Olympics," *Adv. Atmos. Sci.*, vol. 37, no. September 2020, pp. 927–932, 2020, doi: 10.1007/s00376-020-0043-5.

[110] Y. Xiang, L. Gou, L. He, S. Xia, and W. Wang, "A SVR–ANN combined model based on ensemble EMD for rainfall prediction," *Appl. Soft Comput. J.*, vol. 73, pp. 874–883, 2018, doi: 10.1016/j.asoc.2018.09.018.

[111] L. Xu, N. Chen, X. Zhang, Z. Chen, C. Hu, and C. Wang, "Improving the North American multi-model ensemble (NMME) precipitation forecasts at local areas using wavelet and machine learning," *Clim. Dyn.*, vol. 53, no. 1–2, pp. 601–615, 2019, doi: 10.1007/s00382-018-04605-z.

123

[112] H. Yu, G. Wen, J. Gan, W. Zheng, and C. Lei, "Self-paced Learning for K-means Clustering Algorithm," *Pattern Recognit. Lett.*, vol. 132, pp. 69–75, 2020, doi: 10.1016/j.patrec.2018.08.028.

[113] S. Zainudin, D. S. Jasim, and A. A. Bakar, "Comparative analysis of data mining techniques for malaysian rainfall prediction," *Int. J. Adv. Sci. Eng. Inf. Technol.*, vol. 6, no. 6, pp. 1148–1153, 2016, doi: 10.18517/ijaseit.6.6.1487.

[114] N. W. Zamani and S. S. M. Khairi, "A comparative study on data mining techniques for rainfall prediction in Subang," *AIP Conf. Proc.*, vol. 2013, 2018, doi: 10.1063/1.5054241.

[115] Y. Zhou, N. Zhou, L. Gong, and M. Jiang, "Prediction of photovoltaic power output based on similar day analysis, genetic algorithm and extreme learning machine," *Energy*, vol. 204, 2020, doi: 10.1016/j.energy.2020.117894.

[116] L. Zhu and P. Aguilera, "Evaluating Variations in Tropical Cyclone Precipitation in Eastern Mexico Using Machine Learning Techniques," *J. Geophys. Res. Atmos.*, vol. 126, no. 7, 2021, doi: 10.1029/2021JD034604.

[117] https://www.sciencedaily.com/terms/weather_forecasting.htm

[118] http://www.fao.org/3/x5560E/x5560e02.htm

[119] https://www.internetgeography.net/topics/what-is-convectional-rainfall

124